History of Sandstone Orleans County NY

~ by Jim Friday

LIBRARY OF CONGRESS
SURPLUS DUPLICATE

IngramSpark®
La Vergne, Tenn

~ Dedication ~

This book is dedicated to all the hard working immigrants who left their homelands, came to America and labored in the sandstone quarries of Orleans County, NY.

This book is also dedicated to my children and grandchildren in hopes it will help them better appreciate their ancestors and their heritage.

I would like to thank my wife Cary MacDougal Friday for her time and efforts in editing this book and for her unwavering love.

`copyright © by Jim Friday 2021 (All rights reserved.)
www.JimFriday.com

Published by **IngramSpark®**
1 Ingram Blvd.   La Vergne, TN 37086
www.ingramspark.com

Hard Cover - ISBN: 9781087942827
Soft Cover   ISBN: 9781087942520
Library of Congress Control Number: 2021901672

~ Contents ~

	Page
Foreword	4
Introduction	5
Chapter 1: Sandstone Geology	6
Chapter 2: Sandstone Quarry Business	16
◊ 1836 to 1930's	16
◊ Sandstone Products	23
◊ Sandstone Distribution	28
Chapter 3: Sandstone Quarry Methods	35
Chapter 4: Sandstone Workforce	62
Chapter 5: Sandstone Heritage	72
References	95

~ Foreword ~

I cannot recall the exact date I met Jim. Perhaps I was fresh out of library school or had just started working at the Cobblestone Society in Albion as their museum director, but it was well before my tenure as Orleans County Historian. What I do recall is sitting around the kitchen table at the home of Dale Blissett on Linwood Avenue in Albion. The three of us gathered to discuss our shared descendance from Matthew and Hedwig Kaniecki, our family's history and a desire to identify our ancestors' shared place of origin. Over the years, I have had the pleasure of collaborating with Jim on several projects including efforts to transcribe and index sacramental records from St. Mary's Assumption Church in Albion.

The **History of Sandstone in Orleans County NY** helps shed contextual light on the history of the sandstone industry in Orleans County. Since the industry. as our community knew it, ended after World War One, local historians have taken a piecemeal approach at writing about sandstone history. Other writings approached it from a limited geographical or time perspective. Jim's work provides a broad overview of the geological, economic, and social aspects of the sandstone industry from a county-wide perspective.

This book calls attention to the ways in which the sandstone industry of Orleans County added to the diversity of our communities. It focuses on the growth and development that quarries provided for settlements along the Erie Canal, but does so in a way that does not forget the hazardous conditions that quarry labor created. Jim has done an excellent job of assembling both vintage and recent photographs to help tell this important story.

   by **Matthew Ballard**
- Former Orleans County Historian (2015-2020)
- Assistant Director for Collections Strategies at Davidson College
- Curator of www.AlbionPolonia.com

~ Introduction ~

My parents were born and grew up in Albion, a small town in Western New York. My paternal grandparents were born in what is now Poland and immigrated to this country in the early 1880's. My maternal grandparents were born in America shortly after their parents immigrated from Poland in 1882. My mother and maternal grandmother were the historians of our family and that role was passed down to me.

This book was a direct result of my genealogical research, and influenced by my background and personal interests. I graduated college with a degree in chemistry and went to work for the Eastman Kodak company in Rochester, NY. Most of my 34 years with Kodak were spent in their research laboratories designing, developing and commercializing photographic products. As a result of those experiences I developed a keen interest in science, technology and photography.

I grew up in Batavia, NY, just 18 miles south of Albion. We visited my grandparents almost every weekend. My maternal grandparents owned a small dairy farm along the Erie Canal in the village of Albion where I spent a good portion of my summer vacations. I doubt if I was much help on the farm, but I have very fond memories of that time in my life.

Some of my fondest memories were of the local swimming hole on hot summer days where my brother and I would often meet up with a hoard of our cousins and relatives. My "old swimming hole" consisted of a couple of long abandoned, now flooded sandstone quarries. It was not until I got into researching my family history when I realized that my grandfathers, great-grandfathers and many of the other male relatives of my extended family worked in these quarries. This book was my attempt to learn more about how the quarries came to be, how they operated, what they produced, the heritage they left behind and how important they were to my family and the history of Orleans County. Writing this book was also a productive way to pass the time during my weeks of social distancing throughout the Covid-19 Pandemic. Numerous contemporary photos, newspaper articles, academic papers and my personal images were used in preparing this publication. The blue brackets [ ] appearing throughout the text point to the list of references at the end of this book.

Chapter 1: Sandstone Geology

As the title of this book implies, the primary subject is sandstone. It is a type of sedimentary rock formed from bits and pieces of other rocks. For those with an aversion to science, that may be all you wish to know about the geology of sandstone and you might want to skip to the next chapter.

For those that wish to dig a bit deeper, the rest of this chapter covers the backstory of how, when and where Medina Sandstone was formed. It discusses the forces and factors that produced the stone which played a pivotal role in the lives and economy of Orleans County, NY.

First a bit of basic geology. 71% of the surface of our planet is covered with water. Only 29% of the surface of our planet is covered by land masses composed mostly of various rocks. As you may recall from grammar school, there are three general classes of rock: igneous, metamorphic and sedimentary. **Igneous** rocks are formed through the cooling and solidification of magma. **Metamorphic** rocks form when preexisting rock types are transformed with immense heat and pressure into other rock forms. **Sedimentary** rocks are formed from deposits of organic matter or pieces of rocks. They fall into three broad categories, namely organic, chemical and clastic. Coal is an example of decayed layers of organic material that have hardened into rock. Materials like limestone, chalk and rock salt are examples of chemical sedimentary rocks formed when minerals were dissolved in water and redeposited through precipitation or evaporation. Clastic rocks, like our Medina sandstone, form from fragments of older rocks. [1-A,B&C]

The relentless forces of wind, rain, earthquakes, grinding glaciers and seasonal freeze/thaw cycles can erode and break large rocks into smaller bits called aggregate. Most clastic sedimentary rocks form when the aggregate comes in contact with moving water. Currents and wave action move and sort the sediment. Large rock fragments are the first to settle to the bottom while smaller fragments get carried away. At some point the smaller aggregates also fall to the ocean floor. Clastic sedimentary rocks are often classified according to the particle size of the aggregate (see table below).

Loose Aggregate	Size Range	Consolidated & Cemented Rock
Clay	Less than 0.004 mm	Clay, Mudstone and Shale
Silt	0.004 to 0.06 mm	Siltstone
Sand	0.06 to 2 mm	Sandstone
Pebbles	2 to 64 mm	Gravel

Over time, additional deposits covered and compressed the older layers of sediment. As more time passes, small quantities of chemicals dissolved in water often seeped into the sediment, bonded with the grains of aggregate and fused them into a solid mass. Solubilized silicate, carbonates and iron oxide minerals were common cementing agents in sedimentary rocks.[2A&B]

The graphic below illustrates the overall composition of the upper crust of our planet. About 73% of the land surface of our planet is made up of sedimentary rocks. However, sedimentary rock accounts for only a small portion (8%) of the total composition of the upper crust of our planet. The remainder of the upper crust is made up of 27% metamorphic and 65% igneous rocks. [3-A&B] About 62% of the upper crust of the earth is composed of rocks containing various forms of the mineral

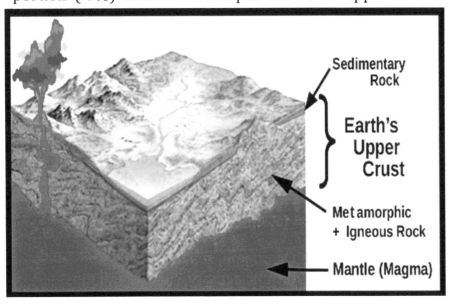

silicone dioxide (SiO2), also known as quartz.[4] Most rocks initially form deep within the crust of the earth and some get brought to the surface through geological forces. Once at the surface, environmental forces attack the rocks and begin breaking them down. Much of the common beach sand found around the world is composed of bits of quartz originating from former subterranean rocks.

Now for more details specific to Medina sandstone. 440 to 418 million years ago was a time geologists call the Silurian period. As the Silurian period began, the planet was emerging from a severe ice age that had killed off most life forms.[5-A] Global temperatures as well as sea levels rose dramatically. Most of what was to become the North America continent was covered with a warm shallow sea. In the coastal areas of the Silurian seas, ocean currents transported the finest rock particles created over millions of years and deposited them in deep beds. It was during the 22 million years of the Silurian period that the sediment that became Medina sandstone was deposited. An analysis of local sandstone reveals it is composed of about 95% quartz. The fine grains of aggregate are semi-rounded and subangular measuring between 0.05 and 0.5 millimeters in diameter. [5-B]

In its purest state, quartz is a very hard, colorless mineral. Trace amounts of other minerals are what give sandstone its color.[6] There are natural variations in the color of the sedimentary deposits quarried across Orleans County and at times even within the same quarry.

Color Variations of Medina Sandstone
St Mary's Church - Albion, NY

The quarries around Medina, NY produced mainly a white or light gray stone with wisps of red, yellow and green. The grains were cemented predominantly with silica and small amounts of lime. Most of the stone produced east of Medina, around Albion, Hulberton and Holley, NY, had a distinct red or reddish brown hue attributed to the high level of iron oxide cementing material. [7-A,B&C]

Following the deposition of the Silurian sediment, sea levels rose and fell again, there were periods of frequent volcanic activity, continental masses collided, mountains were pushed up and glaciers acted to wear them down.

The geological theory of plate tectonics describes how parts of the upper crust of earth move over its interior molten structure (mantle).[8] The theory explains how the continents, mountain ranges and oceans were formed and evolved over time.

Between 330 to 250 million years ago land masses that were to become Europe and Africa began colliding and fusing with part of a land mass that was to become North America. Over millions of years, vast sections of the crust of the earth were thrust upward, folded, deformed and compressed. [9-A,B,C&D] The figure on the top of page 10 [9-E] shows the various segments of the Appalachian Range that resulted from the collisions and which now defines much of the topography of the eastern coast of North America. The lower illustration on page 10 [9-F] shows an elevation profile of a slice of earth on a line running from the shores of the Atlantic Ocean north west to Lake Ontario. When the Appalachian Range was formed, it was likely as high as the much younger Himalayas. Over time, weather and glacial activity have worn down the Appalachians.

The figure on page 11 shows a vertical slice of the bedrock of Western, NY from the Pennsylvanian boarder to north of Lake Ontario. It shows the southern layers of bedrock were pushed down slightly when the Appalachian range was formed.[10-A&B] At the same time, the northern edge of bedrock layers were slightly tilted upward. Glacial activity later eroded much of the top layers of soil and rock, often referred to as overburden, exposing the underlying Silurian sedimentary rock.

The drawing on page 12 shows deposits from the Silurian period run east to west across Orleans County in a band about 8 miles from the south shore of Lake Ontario.[11-A] The sandstone in Orleans County is part of several distinct layers of sediment deposited during the Silurian period. Collectively those deposits are referred to as the "Medina Group".[11-B] Only the oldest sediment deposited during the Silurian period produced rock of suitable quality for building and is referred to as Medina Sandstone. The commercial stone lay in a very narrow band only about 1 mile wide running along the path of the Erie Canal (yellow band in drawing on page 12).[11-C,D&E] In Orleans County, Medina sandstone lays very close to the surface most of the time, but to the east and west, it lies much deeper. The youngest, top layers of the Medina Group are composed of Silurian shale and are usually not commercially useful stone. Older layers of rock from the Ordovician Period found below the Medina Group are also mostly shale.

Formation of Appalachian Mountains
~ Convergence of Continental Plates ~

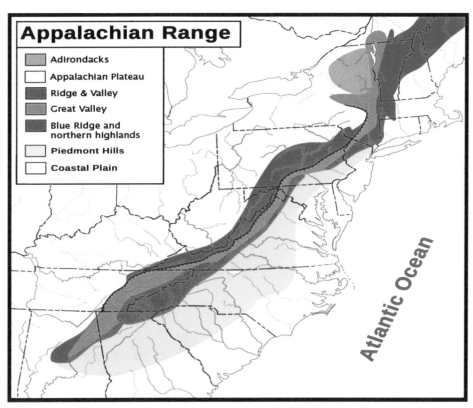

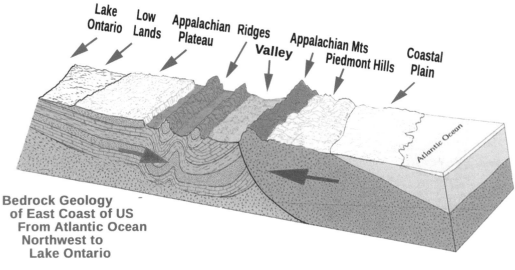

Bedrock Geology of East Coast of US From Atlantic Ocean Northwest to Lake Ontario

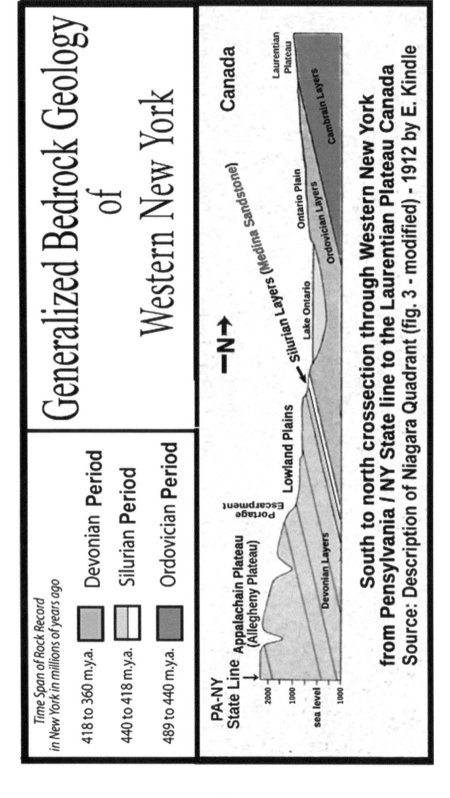

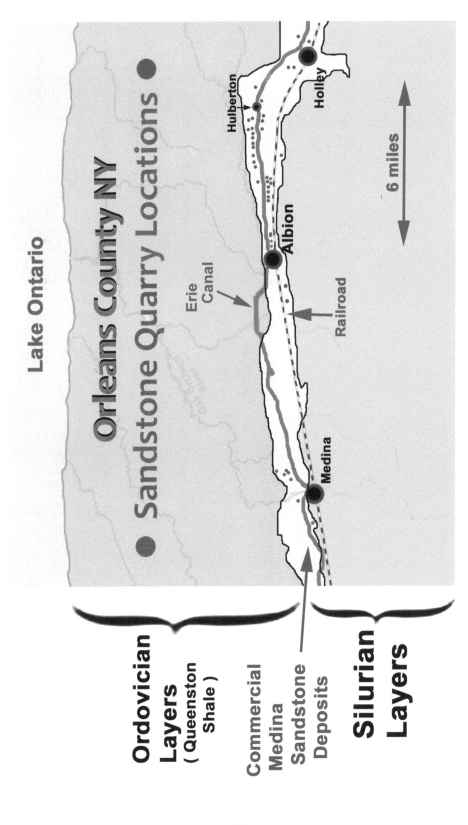

The photo below was taken at the Lower Falls of the Genesee River gorge in Rochester, NY. It shows the stratification or bedding planes of some of the deposits laid down during the Silurian period. Notice a layer of shale and white Medina sandstone are positioned above the red Medina sandstone deposits. Some commercial quarrying of red Medina sandstone did take place below the lower falls in Rochester, NY at Hanford Landing. Geologists at the local university have thoroughly studied and documented the rock formations within gorge.[12-A&B]

Exposed Layers of Silurian sedimentary rock at Lower Falls of Genesee River gorge in Rochester, NY

Over eons of time the natural processes that compressed, tilted and deformed the bedrock of Western New York partially fractured the parallel bedding plains of Medina sandstone. The fractures (joints) run perpendicular to the bedding plains forming natural blocks, which actually facilitated the harvesting of the stone.[13-A&B] The bedding plains and joints of Medina sandstone can be observed in the Genesee River gorge.

Peck's quarry, located east of Albion, off Keitel Rd adjacent to the canal also shows the layer structure typical of many sandstone beds in Orleans County (image below).

Peck's Quarry ~ one of my old swimming holes

The Medina Group has received much attention of late from the petroleum industry. Shale and stone from the Medina group of the Silurian era and younger deposits from the Dovonian era contain small voids between the aggregate grains which often harbor stores of natural gas.[14-A,B&C] The deposits are found in what is called the Marcellus Shale Formation which is located throughout the Appalachian Plateau region (see map on page 10). There are a number of active gas wells in West Virginia, Pennsylvania and southern New York.

The photo below shows a fine example of the application of both white and red Medina sandstone quarried in Orleans County.

First Baptist Church – Batavia, NY

Chapter 2: Sandstone Quarry Business

This second chapter deals with the sandstone quarry business in Orleans county, from its humble beginnings, through its rise to become one of the major industries in western New York and finally to its decline. Along the way, the types of goods manufactured and how those products made their way to customers will be discussed.

◊ 1836 to 1930's

Probably no one was more instrumental to the sandstone industry in Orleans County than **John Ryan**. He was born in Philadelphia, Pa. in February of 1801. His family originally came from Ireland and had been in this country for a number of years.[15-A] His father was a veteran of the Revolutionary War and had discharge papers signed by none other than George Washington. John was trained as a master mason and after arriving in Medina in 1825, proceeded to apply his trade throughout the area. In 1836, New York State approved plans to enlarge the Erie Canal. John Ryan was hired as the contractor to replace all the bridge abutments along the canal between Lockport and Albion, NY.[16]

About 1823, during the construction of the first generation of the Erie Canal, large deposits of sandstone were encountered at Oak Orchard Creek near Medina, NY. The stone was found to be useful as building material. In order to complete his contract to replace the canal bridge abutments in 1836, Ryan needed a good supply of stone. He opened a sandstone quarry in the village of Medina just north of the canal. A marker at the corner of Gravel and Ryan Roads identifies the location of the quarry. Ryan later served several terms as superintendent of the western region of the Erie Canal.[17] He also continued to operate his quarry and supplied stone to meet the growing demand created by the influx of new settlers into Western, NY. The proximity of his quarry close to the canal made it easy and economical to transport his products to markets throughout the country. Shortly after John Ryan opened his quarry, additional commercial deposits of sandstone were developed along the canal east of Medina, NY near the villages of Albion, Hulberton and Holley.

By 1890 the quarry industry had grown to the second largest sector of the Orleans County economy behind agriculture. The number of quarries had increased to at least 27, employing several hundred workers. The table on the next page lists the quarry owners, their relative locations, and the markets they serviced.[18]   Two additional references from 1882 and 1894 provided extensive lists of quarries, owners, their product output and customers. [11-C&D] The quarries were generally small, independent, family owned operations each employing a few men up to as many as 200. Ownership of the quarries frequently changed hands, and operations at some sites were sporadic. At the height of the industry about 50 quarries were opened in Orleans County employing as many as 2000 men.[11-E]

The red dots in the figure on page 12 show the approximate locations of many of the now dormant quarries in Orleans County. It is relatively easy to spot the regular shapes of the quarry ponds clustered along the Erie Canal and railroad line by examining satellite imagery. Websites like Google Earth now provide free satellite views of almost any place on earth.

As the country moved into the 20^{Th} century, the sandstone industry in Orleans County was at its zenith. Large cities such as New York, Buffalo, Rochester, Cleveland, Detroit and Chicago had an ever increasing demand for stone products.

However, it appears many of the quarries were not especially profitable. Most were inefficient operations using antiquated mining techniques. The sheer number of quarries in the county presented a problem in that they were all in direct competition with each other. As a group they lacked a coherent marketing strategy for their products. The 1899 "Medina Sandstone Quarry Prospectus" summarized the market potential and rational for consolidating the many quarries in Orleans County into one unified operation.[7-B]

Orleans County NY Sandstone Quarries (1890)

Owners	Town	Color	Products / Market
Downes & Gorman O'Brian & Co. Fletcher & Sons Michael Slack Big 6 Stone Co.	Holley (near canal & RR)	Light red fine grain	pavers, curbing, gutter-stone Roc. Buf. Syr. as far west as Kansan City
Sturaker Sullivan T. Lardner R. O'Reilly A. Squire L. Cornell C. Vall York C.S. Gwyn Scanlon Hebner Brothers George Hebner E. Fairhen Ford	Hulberton (west and N of canal)	dark red/ fine grain	large block, building blocks paviers, curbstone Roc, Buf & western cities
Goodrich & Clark Co. Albion Stone Co. Gilbert Brady	Albion Multiple quarries east of town betw canal/RR	light red/ fine grain (joints) 200 men	pavers, curbing, gutter-stone sidewalks, block, riprap Erie, Akron, Clevland Toledo, Columbus, Detroit, Chicago, Milwaukee
Kearney & Barrett A Holloway Sara Horan Buffalo Paving C.A. Gorman	Medina (N&E of town)	light gray and some varigate with red, light red & white. very hard	building stone, pavers
Noble & Lyle	Medina (far NE)	redish brown, soft	Block, pavers, rick-rack Roc. Buf. Syr. Erie, Toledo, Columbus, Detroiy, Milwaukee, Omaha, Kansas City

New York State Museum V. 2. No.10, Building Stone in New York, - J. Smock (1890), p260-265

Second only to John Ryan, **Bird Sim Coler** was possibly the most renowned, some might say infamous, figure in the development of the sandstone industry in Orleans County. Coler was an investment broker in New York City and a powerful municipal and state politician. He served as the first Comptroller of Greater New York from 1897 to 1901. In 1902, he was the Democratic nominee for Governor of New York, but lost. In 1905 he was elected to a four year term as President of the Borough of Brooklyn, NY.[19-A] Coler was also the driving force and a prime financier behind the 1902 formation of the **Medina Quarry Company**, often referred to as the "syndicate". Twenty seven of the independently operated sandstone quarries in Orleans County were consolidated into one unified company employing over 1200. Only about six owners of active quarries in Orleans County decided to remain independent and not participate in the consolidation.[19-B,&C]

Bird Coler served as the first president of the new Medina Quarry Co. Most of the remaining board members of the company were associates and close friends of Coler. When the quarry owners agreed to sell their properties to the syndicate, they agreed to receive half of the payment in cash and the remaining half in the form of interest bearing corporate bonds. One of the first acts of the company was to issue $1,200,000 of bonds to compensate the former quarry owners for their properties and to secure mortgages. The board also authorized the sale of shares of common stock worth $2 million. Coler and a number of other close associates purchased some of the bonds and received over half of the voting shares of common stock as a bonus.[19-D&E] In effect, a very small number of investors controlled the sandstone industry in Orleans County.

The capital generated by the consolidation of the quarries was intended to vastly expand and modernize operations. The increased quarry capacity allowed the syndicate to take on projects that would require quick delivery of large quantities of stone. The new company also employed marketing and sales agents to promote sandstone products throughout the country.

During the first few years of operation the syndicate secured several large contracts for paving stone. Output and revenues doubled and it appeared the Medina Quarry Company was doing very well. However, operating expenses were high and they often lost contracts to the independent, unaffiliated local quarries. By 1906 the syndicate suddenly declared bankruptcy and on paper was $110,000 in debt.

 Bird Coler and his partners purchased the outstanding bonds of the bankrupt company for two cents on the dollar. In the process, most investors holding bonds and shares of common stock, lost huge sums of money. Cole helped form a new firm, the **Orleans County Quarry Co**. [20-A] As a result, Bird Coler and his partners not only controlled, but also owned a majority of the quarries in the county.

It was later alleged the bankruptcy of the Medina Quarry Company was a carefully planned strategy intended to gain ownership of all quarry assets. [19-E,F,G&H] In 1908, a court ruled that a $1,000,000 lease and bill of sale of assets of the Medina Sandstone Co. to the Orleans County Quarry Co. was improper and invalid.[19-I] The inference was the assets of the bankrupt company were fraudulently looted.

If the actions of Bird Coler sound wholly improper, matters got even more sordid after the bankruptcy, when he became President of the Borough of Brooklyn, NY. The four years Cole held that position were riddled with numerous, very public scandals. As President of Brooklyn, Cole was responsible for assigning and approving all public work municipal contracts. In 1908 an expose' by the Brooklyn Daily Eagle newspaper alleged Cole lobbied for and approved large paving contracts in Brooklyn requiring the use of Medina sandstone. It was also alleged the Medina Sandstone Company, in which Cole continued to maintain a financial interest, had a virtual monopoly on the supply and price of the stone. [19-J&K] Cole sued the Eagle for libel. After 16 months of legal wrangling and court testimony, Cole lost his case as he was not successful in disproving the allegation put forth by the Eagle. [19-L,M,N&O]

It also came to light that a public works official in the City of Buffalo, NY was simultaneously on the payroll of the Orleans County Quarry Company as a commissioned sales agent, a definite conflict of interest.[19-P] Despite what today might be considered unethical, if not illegal business practices, Bird Cole was never actually convicted of any wrong doing. While I can not condone the exploitative behavior of Bird Coler, he deserves credit for helping grow and modernize the sandstone industry in Orleans County. His efforts help make 1903 to 1916 boom years for the quarry business.[21-A,B&C]

 As 1917 approached, two events were to have significantly bad impacts on the quarry business. In 1914 fighting broke out in Europe and ultimately escalated into World War I. The United States entered the war in April 1917. By the time the armistice was signed in Nov 1918, over 4.7 million men and women served in the regular and auxiliary US armed forces. About 2.8 million American soldiers saw duty overseas. There were 53,402 killed in action, 63,114 died due to disease, and another 205,000 were wounded.[22] Almost all operations at the quarries were halted during the war.

 Compounding the problem was an outbreak of a new deadly respiratory illness. It later become know as the Spanish Flu or the 1918 Flu Pandemic. It first appeared in Feb 1918 at an army base in Kansas and lasted until about April 1920. With all the troop movements during the war, the flu quickly spread to Europe. One third of the population of the world was infected by the virus. In the United States, President Wilson downplayed the pandemic for fear of hampering the all important war effort. The American death toll from the 1918 Flu Pandemic (675,000 people) is hard to comprehend. To put it in perspective, deaths from the pandemic exceeded the number of American lives lost during the Revolutionary War, WWI, WWII, Korean War and Vietnam War combined.[23-A,B&C]

 Shifting resources and manpower to the war effort caused a significant drop in the demand for sandstone building products. The 1918 Flu Pandemic and war both contributed to a shortage of experienced labor willing or capable of working in the stone quarries. In Oct 1919 the Orleans County Sandstone Company declared bankruptcy. The quarries and all remaining assets were seized and sold at auction on 21 Jan 1920 to Philip S. Hill.[24-A&B] Hill was a member of the team of New York City investors who participated in the initial consolidation and formation of the Medina Quarry Company.

The decade following WWI has often been referred to as the " roaring twenties". It was a period of general economic prosperity and a time that saw huge social changes in America. Demand for sandstone also increased and the quarry industry in Orleans County saw a bit of a resurgence. Philip Hill began leasing many of the former syndicate holdings under a series of contracts and royalty agreements. Around 1920 a group of local businessmen and former quarry owners stepped forward and reorganized the Orleans County Quarry

Company. By 1922 there were at least twelve quarries operating again in the county. However, after the war, a serious shortage of experienced stonecutters continued to hamper quarry operations.[24-B&C] Across America, changing construction methods, building materials and customer preferences were also putting pressure on the quarry industry in Orleans County.

In 1924 the 10 millionth Model T rolled off the Ford Company assembly line in Detroit. [25-A] America's love affair with the automobile was in full bloom. As ownership of automobiles increased, there was an increased demand for smooth, safe and durable roads. There was also the need to better connect the outlying rural areas of the country with the cities. The US Congress created the Federal Aid Road Act in 1916 which accelerated the expansion of a national network of highways. [25-B] Most of the urban streets in America at that time were still lined with sandstone paving blocks. Block pavement might have been preferred for urban areas during the era of horse drawn vehicles, but it was less than ideal for automobiles. Asphalt had been in limited use on roads for many years, but it was expensive and difficult to install. In 1903, F. J. Warren was issued several patents for an improved "hot mix" asphalt paving material and process which he called "bitulithic". [25-C] The new formulation revolutionized the industry and is basically still in use to this day. The improved asphalt formulation was smoother, more durable, less expensive and easier to install than stone pavers. As the popularity of automobiles increased, city streets and country roads paved with asphalt quickly became the norm and the demand for sandstone blocks rapidly declined.

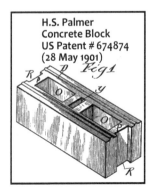

The building and construction industry also saw many changes following WWI. Poured concrete replaced the sandstone flagstone once used for sidewalks. The quarries in Georgia and North Carolina modernized their operations and flooded the market with inexpensive granite blocks and curbstones. H.S. Palmer also received a patent on a method of construction using uniform hollow concrete blocks. They were relatively inexpensive and simple to install. Concrete blocks became mass produced and widely available after WWI. Sears Roebuck & Co. even sold the "Wizard Concrete Block Making Machine" for the do-it-yourself market.[26] Construction using steel to carry the weight of buildings increased in popularity which decreased the demand for structural sandstone blocks. With a steel infrastructure, the outer surface of the building could be clad with less expensive, softer materials like limestone.

Almost all the main markets for sandstone products (road pavers, curbing, sidewalk flagstones, construction blocks) had been replaced or were under pressures from strong competition in the years following WWI. The bottom finally fell out from under the sandstone industry with the crash of the stock market in 1929 and the start of the Great Depression. It was not until the mid to late 1930's that the American economy finally began to recover. By then most of the sandstone quarries in Orleans County had gone under and ceased operations. In 2004 only one sandstone quarry remained open on a limited basis.[27]

The almost 100 year long golden age of the sandstone industry in Orleans County had come to a close.

The demise of our sandstone industry can be attributed to a number of interacting factors. Unscrupulous management, numerous changes in management, competition, changes in building methods, failure to diversify, a world war, a pandemic and a depression all contributed to the passing of an era that was a critical part of the history and economy of the region. Not only were the jobs of quarrymen lost but many railroad employees, canal workers, small contractors, suppliers and people that relied on consumer spending for their wages were negatively impacted when the quarry business declined.

⋄ Sandstone Products

A description of Medina sandstone, published in 1894, appears on the next page.[20-B] It pretty much sums-up the qualities and attributes of the stone.

"....Medina sandstone is characterized by it extreme hardness, compared to similar substances, and by contractors and builders is regarded as one of the best varieties of stone found and utilized in the United States for building purposes. It is devoid of any granite qualities, however, it is easily cut and fashioned into most artistic of architectural shapes and for durability it surpasses the average building stone, the action of climate changes having little influence upon its disintegration. It can be hewed and split, will not crack or break like limestone, and does not wear smooth and slippery like granite. It is usually found in colors of light gray to brownish red. Naturally rich in beauty, it is susceptible of a variety of uses and for architectural purposes."

This section takes a more in-depth look at the stone products that came out of Orleans County quarries. In 1914 the types and relative percentages of sandstone products manufactured in Orleans County were; 49% paving blocks, 36 % curbing and flag stone, 5% building blocks, 6% rip-rap (large chunks of stone) and crushed stone plus 4 % other products.[28]

In the mid to late 1800's, horse and buggy were the main means of transport. Durable sandstone blocks were the preferred material to pave city streets as horses were less likely to slip when the roads were wet. There was a building boom during the mid 1800's through the early 1900's as immigrants flocked to the large metropolitan areas of America. In 1903, Brooklyn, NY purchased 45,000 sq yards of Medina sandstone paving blocks with dimensions of 4 inches thick by 6 inches wide by 7 to 12 inches long, and curb stones with dimensions of 4 inch wide x 18 inch deep and 3 to 6 ft long.[29] An extensive review of Medina sandstone over a period of 50 years and a list of cities that installed it was published in 1912.[30-A] 85% of the stone quarried in Orleans County went to satisfy the demand for street pavers, curbing, gutter stone and sidewalk flagstone.

Medina sandstone sidewalk flags and curbstone are still visible in many places around Orleans county. There is also a strong market for reclaimed stone. [30-B] Most of the street pavers have been removed or covered with asphalt. The one remaining example of exposed street pavers I know of, is Beaver Alley off Main Street in the Village of Albion, NY. Images of reclaimed stone and street pavers appear on the next page.

BEAVER ALLEY   It is funny how sometimes the simplest action can cause long forgotten memories from ones childhood to suddenly pop into your head. After taking the photo of Beaver Alley, I drove through the alley to exit onto Main Street. The washboard ride and old familiar clip-clop-clip-clop sound of the tires on the paving stones brought back fond memories. However, those new Model T owners of the 1920's were not quite as enthusiastic and nostalgic as me about the ride provided by streets paved with Medina sandstone.

Reclaimed Medina Sandstone from Rochester NY

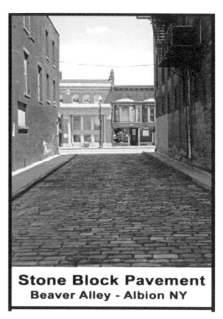

Stone Block Pavement
Beaver Alley - Albion NY

Building blocks were a small, (6%) but lucrative item produced in the quarries of Orleans County. The blocks were usually sawed smooth and square on four or five sides. The exposed front face would be hand chiseled with a decorative pattern. (figure below) The blocks were cut in random sizes and usually laid with thin, tight joints which came to be known as the Ashlar pattern. Ashlar style construction, with a natural stone face, was a popular design frequently used in the construction of churches and large public buildings.

Netural Rock Face   Smooth Face   Pitted Face
Hand Finished Sandstone Blocks

The entrance to the Pullman Church in Albion, NY pictured below is an outstanding example showcasing several sandstone products from Orleans County. The church was constructed entirely with random size red Medina sandstone blocks with a natural stone face and is a fine illustration of Ashlar pattern masonry. Carved red sandstone trim, street curbing, sidewalk flagstones and stone slab steps are also on display.

Entrance to Pullman Memorial Universalist Church – Albion, NY

Crusted stone, gravel and large irregular chunks of stone known as rip-rap, made up about 5 % of the output of the quarries in Orleans County. Much of that stone was used as fill, road bedding and to line the banks of the Erie Canal.

Rip-rap Along Erie Canal Bank Near Brown Street - Albion NY

"Other" stone products made up the last 6% of the output from the Orleans County sandstone quarries. Those products included, in part, such items as window and door lintels, tombstones, large blocks for monuments, bridge abutments, breakwalls, foundations, retaining walls and stair treads. Ornately carved blocks like those found at the Richmond Library in Batavia, NY were high end items also produced by local stone carvers.

Sandstone Carvings
Richmond Library - Batavia NY

◊ Sandstone Distribution

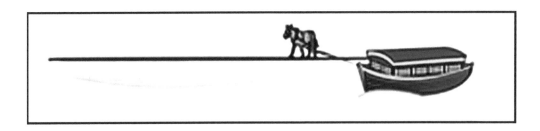

This section explores how the quarry owners of Orleans County were able to deliver their products to their customers. From the earliest days of the sandstone business in Orleans County there was a strong interdependence among the quarries, Erie Canal and the railroads. All three industries thrived about the same time and evolved simultaneously. Virtually all of the quarries were located within a couple of miles of either the canal or rail service. Without an inexpensive, accessible way to transport their heavy products, the quarries would not have been able to expand into distant markets. The canal and railroads also benefited from the quarries by having a ready supply of stone for their construction and maintenance projects.

In the early 1800's, much of the vast western territory of America was sparsely populated. The territories offered a huge financial potential, but movement of settlers and goods into and out of the area involved a long, treacherous and costly journey. Mayor Dewitt Clinton of New York City, a forward thinking politician, recognized a canal from Albany to Buffalo, NY had great potential. It would create a continuous navigable waterway between the major seaport of New York City, Lake Erie and all ports on the western Great Lakes.[31-A&B]   The Erie Canal created the first convenient link between the Atlantic Ocean and the interior heartland of North America.

Construction of the Erie Canal began in 1817 and was completed in 1824. At the time it was the largest public works project in the US and hailed as an engineering marvel. As mentioned earlier, the sandstone discovered near Medina, NY was used as construction materials for the initial canal and ultimately helped launch the quarry business in Orleans County.

 The illustrations below show a map and profile of the first generation of the Erie Canal that was 363 miles long, 40 ft wide and only 4 ft deep. The first boats navigating the canal could carry about 30 tons of cargo. As the photo at the top of the next page shows, the boats were attached to a tow rope and pulled along by mules.[20-C]

Map and Profile of First Generation of Erie Canal

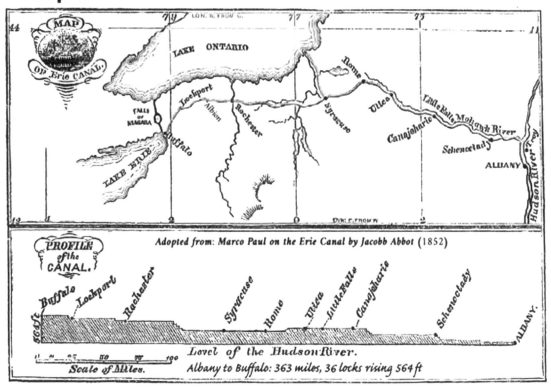

Adopted from: Marco Paul on the Erie Canal by Jacobb Abbot (1852)

Albany to Buffalo: 363 miles, 36 locks rising 564 ft

Before the canal was built it took 15 to 45 days to travel from Albany to Buffalo by wagon and transport of cargo cost about $125 a ton. After the canal was completed, a trip across the state took less than 9 days at a cost of $6 per ton of cargo.[32] As a result, thousands of settlers and tons of goods were transported into and out of the western territories on the Erie Canal in the early years. At its peak over 50,000 people made their livelihood from the canal. The waterway was a resounding commercial success from the moment it was opened. It transformed America, making goods cheaper, more available and even sparked a tourism industry..[33] Profits from travel and trade along the canal also transformed New York City, and New York State into financial giants.

Mule Drawn Toweline Along Early Erie Canal

The success of the Erie Canal united America, spurred economic development and generated a flurry of canal construction. The map to the right shows the network of over 1500 miles of canals in service in America by 1832. [20-I&D]

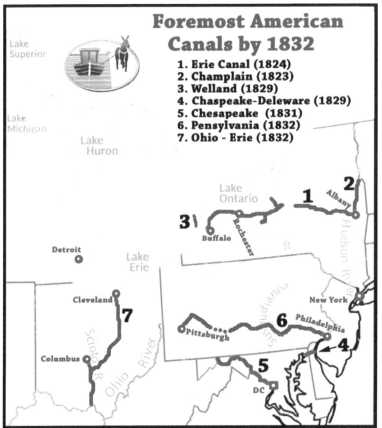

Foremost American Canals by 1832
1. Erie Canal (1824)
2. Champlain (1823)
3. Welland (1829)
4. Chaspeake-Deleware (1829)
5. Chesapeake (1831)
6. Pensylvania (1832)
7. Ohio - Erie (1832)

In 1836 it was decided the canal needed to be enlarged to handle the growing volume of traffic. The canal was expanded to 70 ft. wide and 7 ft. deep allowing boats to carry 240 tons of cargo. Boats on the canal also began to slowly make the transition from mule to steam engine propulsion. [20-E]

Steam Powered Cargo Boat on Erie Canal ~ Albion NY

1836 was also the year the enterprising John Ryan opened the first sandstone quarry in Orleans County. Once the second expansion of the canal was completed, he went on to use it to ship his heavy stone cargo to most of the large cities across New York State and around the great lakes. While New York State was experiencing an expansion of its system of canals, its fledgling railroad network was also undergoing rapid growth. In July 1852 a regional rail line between Rochester and Lockport with connections to Buffalo, NY was opened.[34] It provided Orleans County with frequent passenger and cargo services to points east and west.[20-F]
In 1853 the New York Central Railroad Company purchased and linked 10 regional railroads along the Erie

Train Arrival NY Central Depot - Medina NY

Canal from Albany to Buffalo, NY. The map below shows the company went on to create a rail network connecting the interior of America with the East Coast.[20-G]

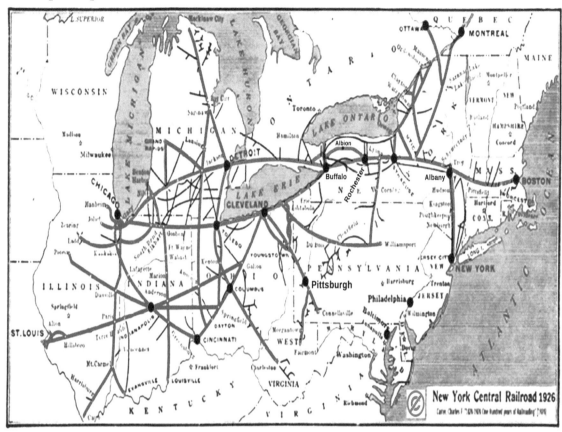

In 1851, before the New York Central Railroad consolidation, it was estimated that it would take six double track rail lines to handle the tonnage of cargo carried on the Erie canal. However by 1883, the technology had advanced and railroads were carrying more cargo than the canal. Rail costs were slightly higher per ton than transport by the canal. However, rail service was faster and ran all year long rather than only seven months on the canal. Travelers also preferred rail transport and passenger ridership on the Erie Canal collapsed.[35]

Between 1905 to 1918, construction took place on a third generation of the Erie Canal in great part to make it more competitive with railroads. The canal was widened to between 120 to 200 ft. and its depth was increased to at least 12 ft. The expansion ushered in the use of large steam and later diesel powered steel barges and tugs. Vessels like the ones pictured below had the capacity to carry 3000 tons of cargo.[20-H, 31-A&B] The various canals around the state were also consolidated to form the New York State Barge Canal Corporation.

Hulberton Sandstone for Buffalo Harbor Breakwall

Competition between the canal and railroads was always fierce and helped keep shipping costs low. Railroad transport also allowed the Orleans County quarry business to expand into markets not accessible by water.

There were multiple examples of friction between the quarry owners and the Canal authority. Most of the problems seemed to have occurred as a result of the last reconstruction and widening of the canal.

The court case of Frank Colonna vs New York State dragged on from 1907 to 1930.[36-A & B] His quarry was located east of Albion, adjacent to the canal. It often flooded, making it difficult and costly to operate. Colonna claimed negligence due to poor maintenance by the canal authority. After several appeals, the claimant won and was awarded damages for lost income. In two similar cases twenty quarry owners in Orleans County also sued the canal authority for flood damage to their properties.[36-C&D]

Another lawsuit filed by the Orleans County Quarry Co. against the State of New York claimed the quarry owner was insufficiently compensated when their properties were seized by the State of New York during the expansion of the Erie Canal.[37] It appeared the claimant was unable to prove his case and lost his bid for additional compensation.

By the middle of the 19th century there were over 3000 miles of canals interconnecting the waterway in the United States.[20-I] However, in the long run, railroads proved much more economical to build and maintain than canals. Only a small number of canals in the United States remain in service today as most were abandoned or have fallen into disrepair. The NY Barge Canal System continues to be maintained. It still carries some commercial traffic, but is now used mainly for recreational purposes.[20-J]

Chapter 3: Sandstone Quarry Methods

While there are some fragmented first hand accounts **[38-A&B]**, I have not located a comprehensive contemporary account of the methods and equipment used to quarry sandstone in Orleans County. However, a number of vintage photos of local quarries have survived. There are also several technical references written in the late 1800's to early 1900's that discuss in detail the quarrying methods used by the sandstone industry in America. **[38-C,D&E]** Additional references written more recently also cover this general topic from a historical perspective.**[39-A&B]** In this chapter, the fragmented accounts, vintage photos and mining references were pieced together to paint as accurate picture as possible of the techniques, workflow and equipment used in our local sandstone quarries.

When an operator was trying to decide if he should make the sizable investment necessary to expand or open a large quarry, he would often drill a series of test holes and extract rock core samples. The data would tell the operator if he had commercial grade stone, how much overburden he would need to remove, plus the thickness and distribution of the deposit. That information would allow the owner to know how much stone the site might yield and help him plan what techniques were needed to efficiently excavate the quarry. The best commercial stone needed to be strong, durable, uniform, appealing and inexpensive to harvest. In Orleans County only a few feet of overburden usually needed to be removed to access the commercially valuable sandstone deposits. Most of the quarries were small relatively shallow open pit operations. The quarry owner would often use explosives to open a breast or wall face where the stone could be worked. A series of holes would be drilled in the rock, charged with black powder, packed with sand and detonated. Blasting generally produced a lot of rip-rap and fill stone. Blasting was as much an art as it was a science.[40]

In the early days, bore holes would have been drilled manually with a chisel and hammer. The image below shows a three man crew of rock-splitters boring a hole with a star drill.[41-A] The center man held the bit and kept rotating it slightly while the other men swung the hammers. On a good day, a three man crew could drill about 40 to 50 linear feet of holes.

Manually Drilling Rocks (*Double Jacking*)

"Plug and Feather" was a low-tech method often used on the quarry floor to split and shape stone.[41-B&C] First a series of holes would be drilled along a line where the stone was to be split. Two thin pieces of steel (feathers) would be placed in each hole followed by a steel wedge (plug). The image below shows a quarryman proceeding down a line of wedges tapping each one in succession with a hammer. He would continue to tap the plugs in sequence until the stone split. With an understanding of the grain of the rock, very large blocks of stone could be shaped or even split off from the breast of the quarry.[41-D]

Plug & Feather Method for Splitting Rocks

Once large chunks of stone were freed from the quarry bed, workmen might rough shape them using tools like **bull and striker hammers**. The technique is shown in the photo at the top of the next page. One man would hold the bull set, moving it along the line where the stone was to be cleaved while a second man would wield the strike hammer.[42-A]

Splitting Stone with Bull & Striker Hammers

Below is a vintage photo of a shallow pit quarry in Orleans county.[42-B] An array of simple tools held by the men and scattered around the yard include quarry hammers, sledge hammers and a T-square. A wedge shaped bull hammer can be seen in the foreground of the photo. It appears the work crew of 15 men were harvesting stone and rough dressing some of the blocks before they were transported from the quarry by horse and wagon.

Over time more labor saving devices were adapted, but a large portion of sandstone in Orleans County would continue to be harvested by hand with rather simple tools.

"**Guy derricks**" like the one shown in the photo shown below, were devices widely used through out the quarry industry.**[42-C]** They consisted of a fixed mast held in a vertical position by securely anchored guy-lines and a boom whose bottom end was hinged to tilt up and down with cables. Some versions allowed the derrick to rotate and its winches could be operated manually or with steam power. Guy derricks with their arrangement of block and tackle were used to move stone around the quarry and load large blocks for shipment. Depending on the size and character of a rough stone block and the customer order, a block would be transported to the sawing shed followed possibly by a trip to the finishing shed where it would be shaped into the desired dimensions.

Guy Derrick

When a rough block arrived at the sawing shed, it would be placed on carts, rolled into place and secured to keep it from moving. The photo on the top of the next page show a couple of views of a **"gang saw"** in a granite quarry powered by a steam engine.**[43-A,B,C,D&E]** Virtually the same equipment and techniques were used in sandstone sawing sheds. A heavy timber frame holding several saw blades would be lowered onto the stone. The blades were made of ¼ inch steel, about 13 ft long by 6 inches wide. They had no teeth and could be adjusted to any width. A steam engine provided the power to draw the heavy saw frame and blades back and forth across the stone. The weight of the frame allowed the blades to cut through the stone. A little sand mixed with steel shot was added to aid the cutting action of the blades, and a stream of water was continually poured onto the stone to reduce heat and carry away the stone dust.

Gang Saw

Once the stone was cut into manageable blocks, if they needed further shaping, they would proceed to the finishing shed. Many of the simple tools shown on the next page would have been in the tool box of most stonecutters and quarrymen.[44-A&B] I still have a couple of my grandfather's sledge hammers and five foot crowbars, remnants of his days in the quarries.

After the formation of the Medina Sandstone Company in 1902, capital was made available to modernize and upgrade some of the quarries. Four portable steam powered winches, several extra large guy derricks and a steam powered channeling machine were acquired. They doubled the size of the stone sawing operation in Eagle Harbor, NY and installed crushers at two locations to convert waste into crushed stone for road material.[45-A] A 1905 article also boasted the quarries were equipped with a variety of steam powered drills, pumps and other modern devices.[45-B]

In the early 1900's when the Orleans County quarries were acquiring new equipment, steam was the predominant power source. Some gasoline, diesel, pneumatic and electric equipment was available, but had not been fully perfected or proven reliable for heavy commercial quarry applications at that time.

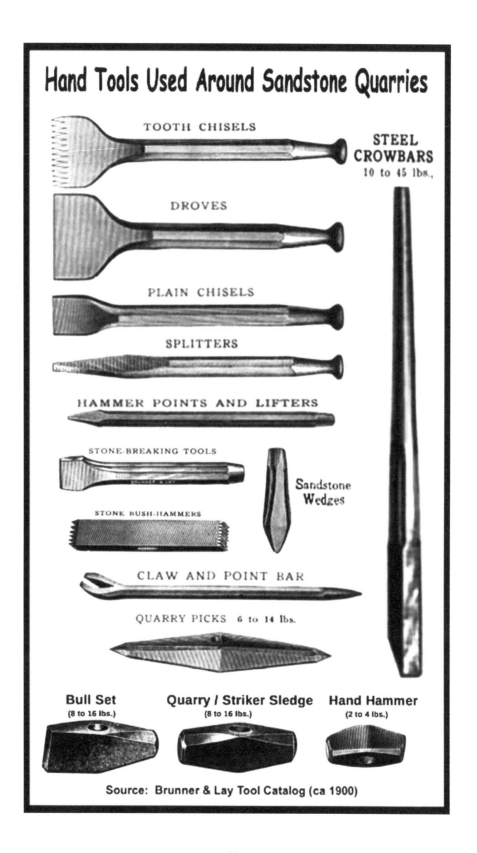

Most of the following vintage photos were all taken at the height of the sandstone industry in Orleans County. The images are presented with the intent to give further incite into the workflow, day to day operations, equipment and environment within the quarries.

One device widely adopted by local quarry operators was the **steam donkey** like the one shown below. It was invented by John Dolbeer in 1882 [patent #256,553] as a portable, integrated machine consisting of a single cylinder steam engine and gears that turned one or more winches containing wire cabling. It was used in combination with a derrick to lift heavy loads. Originally designed as a portable power supply for lifting logs in the woods, it was quickly adopted to lifting heavy stone blocks. The steam donkey revolutionized both the logging and quarry industry.

The image at the top of the next page is a good example of a steam donkey in action.[46-A] The photo was taken slightly east of Medina when a culvert under the Erie Canal was being constructed about 1916. Large blocks of Medina sandstone were being lifted into place with the aid of a guy derrick controlled by the winch of a steam donkey.

Construction of Culvert Under Erie Canal ~ Medina NY (abt 1916)

The image below shows three tall guy derricks in operation in an expansive pit quarry in Albion, NY.[46-B] The first two derricks appear to employ manual winches but a steam donkey was situated near the third derrick. Notice this quarry had about 20 feet of overburden that needed to be removed before the sandstone deposit could be mined.

Albion Sandstone Quarry

The **steam traction engine** like the ones shown in the three photos on the next page was another common machine used in local quarries. It was invented by Thomas Aveling and developed throughout the 1800's.

Its primary purpose was to draw heavy loads behind it, but it could be put to many good uses around the stoneyard. The tractor replaced teams of draft horses and oxen. The first two images on this page were taken at the McCormick Quarry located near the Bates Rd bridge along the Erie Canal in Medina, NY. **[46-C]** On the right side of the top photo, two men can be seen standing on a portion of the bridge.

Steam Traction Engine

In the image below of a "Stone Crusher at Work" **[46-D]**, a steam traction engine is located to the left of the scene. A long looped belt can be seen stretching between the traction engine and crusher unit. One end of the belt was looped around a rotating drum on the side of the steam engine and the other end of the belt was looped around a drive wheel of the rock crusher. In this way, the traction engine was used to power the crusher. My grandfather had an old gas powered, "one-lung" (single cylinder) John Deere tractor on his farm that he would rig in a similar way to power a saw.

The Stone Crusher at Work

44

In the photo on the previous page, the machine that actually crushed the rock was located toward the back of the horse drawn wagon and at the base of the conveyor belt. The device was similar to the one that appeared in a 1879 advertising brochure from the **Blake Crusher** Company shown below.[46-E] The apparatus was invented in 1858 (US Patent 20,542) by Eli W. Blake.

Eli Blake Rock Crusher

It was a heavy duty, reliable unit widely used throughout the ore mining and stone quarry industries to coarsely crush rocks. In a sandstone quarry, the rocks most often sent to the crusher were waste fragments generated by the stonecutters or material judged too small to be make into other products. The diagram on the next page shows the inner workings of the crusher.[46-E] Chunks of stone would be fed into the top of the machine. A movable jaw inside the crushing chamber would travel back and forth pushing the rock against a stationary jaw. Gradually the rock would get pulverized to the point where it was small enough to fall through the gap at the bottom of the crusher chamber. The gap could be adjusted to produce various sizes of crushed stone.

The conveyor belt visible in the photo on the bottom of page 44 was used to transport the broken stone from the crusher up to a trommel which consisted of a rotating cylinder with perforations of various sizes. Crushed rock was fed into one end of the trommel and as it rotated, stone fragments smaller than the perforations would fall through, while larger fragments would flow out the other end.[46-F] The trommel was used to separate the crushed rock into

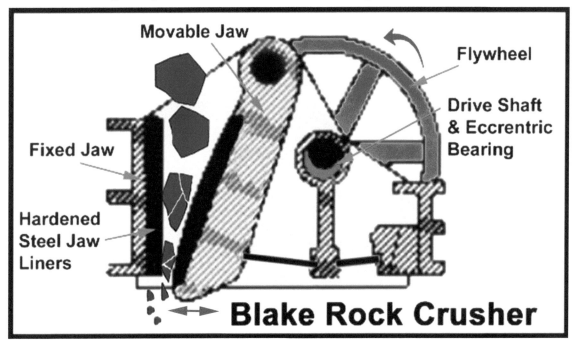

various classes of gravel, ranging from dust to small, medium and large size particles. The various classes of gravel would fall into separate bins in a storage hopper where they could be conveniently transferred into wagons.

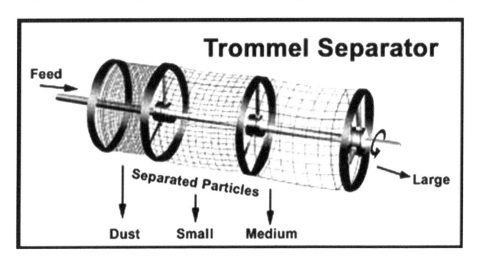

The image below is a group photo of the work crew employed at the stone crusher operation shown on page 44. [46-G] Some of the men are sitting on the rock crushing machine while others are standing in front of the storage hopper. One of the workmen can be seen sitting on the conveyor belt and a rotary trommel is also visible in the upper left hand corner of the photo.

Reports chronicling the 1902 consolidation and modernization of the quarries in Orleans County also listed the purchase of channeling machines.[47A&B] Equipment like the **Sullivan Channeler** was widely used at the time in stone quarries. [47-C,D&E] They rode on metal tracks and had a series of hardened steel blades that would be hammered into the rock creating a channel from several inches to several feet deep. If the remaining sides of the block lay at natural horizontal layers and vertical joints, plug and feather, wedges and crowbars could be used to free the block from the breast of the quarry. Otherwise. a few horizontal and vertical holes would need to be drilled before the block could be moved. Use of channeling machines took more setup time than blasting, but could produce large unbroken blocks of stone with much less rubble and waste. A Sullivan style channeler on rails can be seen toward the middle and back of the Albion quarry photo at the bottom of the next of this page.

Steam Powered Sullivan Channeler

Ingersoll Rock Drill

Boring holes manually in stone for plug and feather or blasting was a tedious and time consuming operation. That job could be accomplished much faster with the introduction of tools like the **Ingersoll Steam Drill.[48-A&B]** One experienced rock splitter could normally drill about 10 linear feet of holes in sandstone per day by hand. An Ingersoll Steam Drill could bore 10 linear feet of holes in one hour.[48-C,D&E] Steam drills significantly increased the speed and reduced the cost of mining. An Ingersoll Steam Drill can be seen at the center of the photo below, in front of the channeler. Bore holes can be seen on the two large blocks in the foreground. The pit quarry pictured below was located around Albion.[49-A] It had a thick, uniform bed capable of producing large blocks of sandstone. There was likely a tall derrick installed in this quarry as a guy line can be seen coming down from the left corner of the photo. After the channel was cut, the bore holes would have been used with wedges to free the block from the bed.

The small image to the left shows a **steam shovel** at work in one of the sandstone quarries in Albion, NY before 1905.[49-B] Similar machines would have offered huge labor savings when it came to stripping overburden, loading rip-rap and removing rubble left behind from dynamite blasts.

The photo on the right also shows a similar behemoth steam shovel used to enlarge the Erie Canal around 1916. [49-C] Notice how a band of rip-rap was installed along the upper edge of the canal bank to stabilize and protect it from erosion.

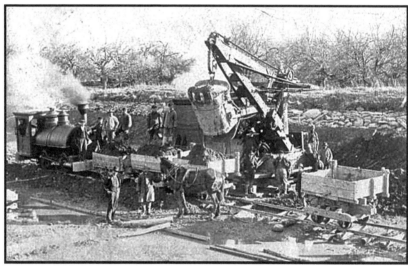

Third Expansion of Erie Canal - Albion NY (abt 1916)

I believe the two images on the next page were taken off East State St. in the Village of Albion, NY. before 1905 at one of the facilities associated with the Orleans County Quarry Company.[49-D] The photos show a sizable, well equipped operation. In the top photo, a **steam powered crane** is visible in the distance. It would have been used to load the stone for shipment or move blocks into the adjacent building which may be the finishing shed. The tall guy derrick and a steam donkey would have been used to move the large blocks piled around the quarry. On the left side of the bottom photo two men were preparing to lift a large stone block from the quarry floor. Drill marks on some of the large blocks give an indication of how they were separated from the quarry bed. The wooden tower at the center of the bottom photo contained a number of pipes that likely supplied water and steam to the lower level of the quarry. A channeler and steam tractor are also visible in the distance at the right edge of the bottom photo.

The image below was taken at the Horan Quarry near Medina about 1904. [49-E] The deposit seems to have been highly layered with distinct vertical joints. Notice the two men at the center of the photo were standing on either side of what appears to be an Ingersoll style steam drill. The trough and tubing to the left of the drill probably held water which would have been used to flush the bore holes and reduce airborne dust.

Below is another old photo from the McCormik quarry.[49-F] The crew of 24 workmen pose with their harvest of sidewalk flagging and curb stone. Child labor was common in the quarries and two members of this work crew looked quite young. As for tools, the young lad seated on the right had a square and most of the other workmen were armed with crowbars.

The photo below was taken about 1900 at the Cornell Quarry located in Fancher, NY.[49-G] The commercial stone was covered with about 15 feet of overburden. The sheer wall of the large pit quarry show a thick uniform bed which was likely the dark redish-brown sandstone typical of the area. In the distance to the left of the photo, a steam tractor and fixed-leg derrick were being used to harvest large blocks of stone, further extending the shear wall of the pit.

The photo below is another image from a typical Hulberton quarry.[49-H] It shows a closer view of a Sulllivan type channeler. Most of the quarries in this area were clustered along the Erie Canal. all within a couple of miles of each other. I suspect an expensive piece of specialized equipment like a channeler would have been shared among the quarries. Also notice there were a couple of children numbered among the work crew.

The image below of about 20 men was taken in another shallow pit quarry around Hulberton.**[49-I]** A few crowbars are the only visible tools. The photo also hints at a common problem faced by the quarry workers. Most of the quarry pits were adjacent to the Erie Canal and below the water table. Often quarries would flood during the annual seasonal shutdown and needed to be drained each spring. Many quarries also needed constant pumping during the season to keep them viable.

The photo at the top of the next page was taken before 1900 at the Farrand Quarry which was located off North Main St. near the Holley Elementary School.**[49-J]** Today, the quarry is hidden in the woods and filled with water. It appears to have had a thick rather uniform bed of sandstone. A pile of large stone blocks and a huge guy derrick are visible in the foreground. The oxen along with quarry workers in the photo give an indication of the massive scale of the operation. Reportedly the largest block of stone produced in Orleans County came from the near by Gorman and Crane quarry. That block measured 36 x 12 x 7 feet in size. **[49-K]**

The photo to the left shows a stone crusher plant once located on the banks of the Erie Canal east of Albion.[49-L] The plant was demolished to make way for an expansion of the waterway. The photo was taken during the winter months while the canal was partially drained and closed for the season. Several conveyor belts were positioned on the side of the building to allow crushed stone to be loaded directly into canal boats. Tracks can be seen leading into the plant allowing delivery of bulk stone and easy off-loading of crushed stone into railroad cars.

The photo below was taken at the old O'Brien Quarry located on Howard Road near the railroad line, just south of the town of Holly, NY.**[49-M]** It was a sprawling complex of several pits that at one time employed several hundred men. Scattered about this quarry are an array of sledge hammers and crowbars. However, there are signs that large machinery had been used in the quarry. There are scratch marks and grooves in the overburden indicating a steam shovel was likely used at some point. Remnants of long bore holes visible on the breast wall were created with a steam drill and used to blast stone from the bed. The quarry was owned by Pasquale DiLaura, who at one point served as President of the Orleans County Quarry Co. He also supplied much of the sandstone for the bridges and structures along the Lake Ontario State Parkway in, NY.

The photo on the next page was also from the O'Brien Quarry. [49-N] The barrel next to the tractor would have supplied water to power the steam engine. This image also provides a fine illustration of the layer structure and mining sequence in many local sandstone quarries. There were two distinct top layers of loose dirt, shale and rubble that needed to be stripped before the commercial stone could be accessed. The upper most layer of good stone behind the men shows remnants of drill holes which were likely used to blast the stone free from the bed. The man wearing the white shirt and bowler hat at the center of the photo was standing on a steam drill. The work crew were all standing on a thick lower layer of stone that appears to have been much more uniform and useful for producing curbstone and large blocks.

The last series of four vintage photos were all taken at the Vincent Quarry located on the north bank of the Erie Canal along Canal Road slightly west of Hulberton.[49-O] Collectively they tell the story of how some curbstones were produced in Orleans County. The first image at the top of the next page offers a closer view of sandstone layer structure within that quarry. Three men were shown standing in front of a deposit of stone containing multiple layers and joints. The men are standing atop two thick layers of uniform stone separated by a thin layer of loose sediment. Those tick uniform deposits appear to split easily by hand and could be used to produce blocks of larger dimensions for curbstone, lintels and stair treads.

The photo below show the same three quarrymen standing on top of a large number of curbstone slabs they produced using their bull and striker hammers.

Medina sandstone like the sample to the left was formed in discrete layers. In Orleans County the sedimentary layers were almost parallel to the bed of the quarries. Depending on the character of the stone, an experienced rock splitter accustomed to swinging a sledge hammer for ten hours a day, could produce smooth faced blocks by cleaving them along their sedimentary layers.

The image below shows more of the curbstone production process. Once a thick uniform layer of stone was exposed, a steam drill was used to bore a line of vertical holes about five feet back from the front edge of the deposit. Large blocks could then be freed from the quarry bed. Notice the top layer of blocks the men were standing on and the orientation of the drill marks. The blocks had been flipped on their sides to reorient the parallel sedimentary layers in a vertical direction making it easier for the quarrymen to cleave the stone.

The last photo below from the Vincent Quarry, taken from a different angle, shows dozens of curbstones cleaved by the three quarrymen. The steam drill used to help free the large blocks from the quarry bed can also be seen leaning against the far wall.

The quarries today are mostly nameless, all but forgotten ponds located off back roads, hidden behind overgrown brushes and scattered around fields of corn stalks and cabbages. Several quarries have been converted to campgrounds, sportsman clubs and recreational facilities. Some of the sites like the 1836 Ryan Quarry in Medina and the sprawling Reed & Allen Quarry off East State in the Village of Albion have been filled in or partially reclaimed.

Most of the quarries are now in private hands and not accessible. The following images are of five quarries visible from the highway or banks of the Erie Canal. They are representative of the 50 odd quarries that once operated throughout Orleans county. West of Hulberton, on the north side of the canal, along the one mile length of Canal Road are a cluster of at least 12 quarries. The sprawling quarry pictured at the top of the next page is littered with old rusting boilers and iron rails that were likely used by a channeler machine.

The image below was taken while I was standing on the towpath of the canal. This quarry is also located on Canal Road directly adjacent to the Vincent Quarry mentioned in the previous pages. The dark redish-brown Medina sandstone ledges and deep, clear water provide a popular unofficial swimming hole for local residents. There are even a couple of well used rope swings for the brave or foolhardy.

The next two quarry photos were taken between Hulberton and Albion along Transit Road at the bridge over the canal. The quarries are situated on the north and south banks of the canal. The picture to the left was taken on the south side of the canal. The towpath is the only thing separating the canal from the quarry pond.

The photo to the left was taken while standing on the north end of the Transit Road bridge. There is a private home adjacent to the quarry. The owner has a dock and appears to be using the pond as his private playground.

The quarry pond to the left is situated adjacent to the south shore of the canal at the Densmore Rd. bridge about one mile west of Transit Rd. There were a couple of sportsmen fishing in the pond when I took this photo. This quarry contained deposits of red, white and yellow variegated Medina sandstone.

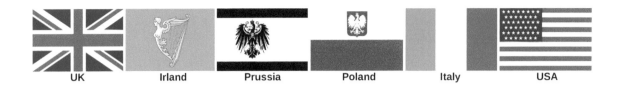

Chapter 4: Sandstone Workforce

This fourth chapter deals with the men who were employed in the quarries. Some of the skills they needed, along with what they earned, the dangers they faced, where they came from and why they came to Orleans County. I used the stories of my immigrant ancestors to illustrate experiences typical of local quarrymen.

There were many different skills needed to efficiently operate a sandstone quarry. Below is a partial list of quarry jobs with a few approximate salaries for a 10 hour day in 1930.**[50-A]** The list is roughly organized in descending order of job skills and salary. At a time when general farm hands were paid about $2 per day and factory workers were paid about $2.60 per day, it is worth noting quarry workers received a premium salary.**[50-B&C]**

Foreman ($10/day) - managed work crew on the quarry floor – usually had a lot of experience with mining procedures and the properties of stone.
Fireman or Engineer ($6/day) – controlled steam donkey, winches and derricks to move large blocks of stone around quarry. My great uncle Felix Siebak was listed on the 1900 and 1905 US census as a quarry fireman.
Certified Blaster ($5/day) - responsible for handling the dynamite and blasting operations around the quarry yards.
Steam shovel and Traction Engine operators
Maintenance Man – charged with sharpening tools and keeping all the equipment around the quarry in good running order.

Stonecutter – trimmed, carved and finished stone, usually with hand tools. A unionized, skilled laborer, often required an apprenticeship ($4+/day).
Rock Splitter or Block Breaker - specialist in splitting blocks from the stone face utilizing plug & feather, steam drills and channeling machines.
Stone Saw Operator - worked in the sawing shed to cut large blocks into usable dimensions for the stone cutter or direct shipment.
Experienced Quarryman ($4/day – 10 hr) – did much of the harvesting of stone, heavy lifting and general work around the quarry.
Teamster ($3/day for driver with horse or ox team) - transported stone.

Work in the quarries was dusty, dirty, backbreaking and often dangerous. There were numerous reports of injuries from falling rocks, careless use of heavy machinery, lifting accidents and dynamite mishaps. Periodically New York State conducted safety inspections of the local quarries. The most frequent infractions sited included issues with storage and handling of explosives, boiler maintenance and child labor violations.[51-A] Dust from Medina sandstone, composed mostly of silicone dioxide, could damage lungs over time. Quarry workers in Orleans County were frequently afflicted with respiratory diseases like silicosis which made them susceptible to pulmonary tuberculosis.[51-B] Ignatius Rice, my great grandfather and a longtime quarryman, died of tuberculosis. Given the difficulty and dangers associated with the job and the specialized skills needed, it is easy to see why quarry workers were paid a premium salary.

Most employment positions in the quarry were seasonal. Depending on the weather, activities began in April and ended in October.[52] A maintenance man might be kept on over winter to repair and prepare the equipment for the upcoming season. Off season was also a good time to strip overburden from new sections of the quarry. A steamshovel operator and a few laborers would be needed for that job. However, most quarry laborers were furloughed over the winter months. Many quarry workers sought temporary employment during the off season to supplement their income. There were plenty of seasonal jobs available. Orleans County produced a wide variety of agricultural products (apples, tomatoes, cabbage, potatoes, onions, cherries, peas, beans, corn, etc). Fall harvesting, pruning apple trees, preparing the fields for spring planting, shipping and preserving the produce all required extra part time labor. There were also cold storage facilities, a dehydration plant, canning factories, vinegar works and dairy farms through out the county, all in need of seasonal help.

There were also a variety of off season jobs available for men who had experience working with stone. In the four years between 1892 to 1896 alone, three large churches were constructed in Albion using local sandstone (Pullman, St. Mary's & St. Joseph's). Much of the stone work and maintenance on the Erie Canal was also done during the winter months while the canal was drained and out of service.

The story of the labor force in the sandstone quarries of Orleans County, like much of the United States, was really a story about immigrants.[52 & 21-B] A survey of the 1880 US census from the Village of Albion provides a crude estimate of the composition of the quarry workforce in Orleans County at that time. Almost all the quarrymen were immigrants. They included 46 laborers from Ireland, 9 from England, 2 from Germany and 5 were of Polish descent.[53] There were also a few additional German immigrants in Albion employed in other skilled professions such a masons and marble cutters

From 1820 to 1920 almost 45 million people from around the world immigrated to America. During the first half of the 1800's the Irish and Germans were the two largest groups to immigrant to the United States. The Irish came to America to escape civil unrest, severe unemployment, and the widespread famine that gripped their country. [54-A&B] From 1845 to 1850 a potato blight further devastated the already deeply impoverished people of Ireland. Between 1840 to 1900, the population of Ireland dropped from 8.2 to 4.3 million with the majority (3.4 million) immigrating to America.

Between 1850 to 1900 about 4.5 million immigrants left Germany and came to the United States to escape crop failures, unemployment, government restrictions, wars and the threat of military service.[54-C] As a general rule, Germans immigrants were more skilled and had

more financial resources than the Irish. Many of the Irish and German immigrants provided the labor to construct the Erie Canal in New York State and railroad networks across the United States.

The Poles were part of the next big wave of immigrants. At one time the Polish-Lithuanian empire was a dominant political and economic power in Europe. By 1795 the empire ceased to exist as all of its territories were divided among the rising neighboring empires of Prussia (Germany), Russia and Austria. [55-A] In 1871 Chancellor Bismarck initiated a series of policies aimed at restricting the political influence of the Catholic Church in the Prussian controlled sector of the former Polish Empire. Some Catholic clergy were expelled from the country, limitations were placed on the ownership of certain personal property and the Polish language was not allowed to be used in schools. Most ethnic Poles were Catholic and saw these policies as an effort to stamp out their culture and they began emigrating in large numbers. About the same time steam powered ocean liners were coming into wide use. An Atlantic crossing from Europe to America could be made in less than two weeks.

The Prussian officials, who controlled the western part of Poland, were only too happy to allow the ethnic Poles to vacate their homeland and made quite a business of exporting people to America. The northeast portion of Poland was controlled by Russia. Austria controlled the province of Galicia in southeast Poland. Both areas were also severely depressed during the late 1800's. Their economies were predominantly agrarian with almost no industry. Farming methods were primitive, crop yields were low and there was widespread famine and disease. Land prices were high and overpopulation became a problem as many Poles from Prussia immigrated to Galicia. The industrial revolution brought some modernization to farming methods in Poland, but that only put even more people out of work. With no industry to adsorb the idle workforce, unemployment became a serious problem. The Austrians and Russians also viewed the excess population as a good source to fill the ranks of their armies. Young men were at risk of being drafted into a long, harsh life of military service. Conditions were so dire that many of the Poles from the Austrian and Russian sectors also decided to abandon their ancestral homeland.

From about 1870 to WWI, it is estimated that about 2 million ethnic Poles, including my ancestors, came to America in search of more religious freedom and more economic opportunities.[55-B] The exact number is difficult to determine because when Poles entered this country or were documented on census records, they were often listed as German, Austrian or Russian.

The next big wave of immigrants to arrive in America and settle in Orleans County came from Italy. One million Italians came to America between 1880 and 1900 and another 3 million arrived between 1900 to 1915. The majority were poor laborers from southern Italy and Sicily. The government in Italy began to repress freedom of expression and the country was gripped by poverty. There were more economic opportunities in America and many came with the idea of earning enough money to return to Italy and purchase a plot of land.[56]

A survey of the 1880 census records shows there were no individuals of Italian descent living in the Village Albion and only nine were listed on the 1892 census. To entice more of their countrymen to immigrate, the quarry owners provided housing in a tract situated on McKinstry Street in Albion. [57-A&B] By 1915 the Italian population in the Village of Albion climbed to 790 people and many of the men were employed as quarrymen. By the 1920's some of the Italian immigrants were managing several of the quarries in Orleans County.[21-B & 24B]   Attracted by the numerous quarries around Hulberton, NY, many of the Italian immigrants eventually settled in that area.

The following stories of how my ancestors became Americans are typical of millions of others immigrants who arrived here over a hundred years ago. Six of the males on my family tree fled their Polish homeland and found their way to Orleans County in search of a better life.[15-B&C]   On the next page is a map of Poland showing the location of their hometowns. Upon arriving in Albion, the local quarries served as their stepping stone to success. With hard work and dedication they went on to purchase homes, raise families and helped make America a better place.

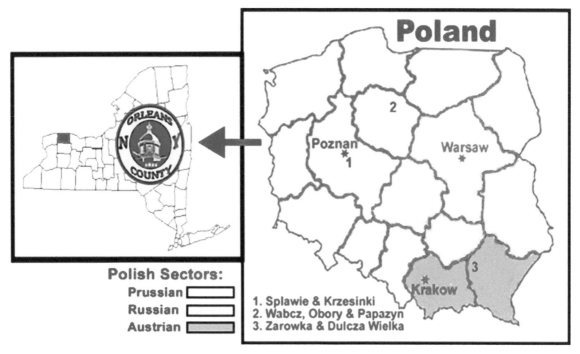

Polish Sectors:
Prussian
Russian
Austrian

1. Splawie & Krzesinki
2. Wabcz, Obory & Papazyn
3. Zarowka & Dulcza Wielka

John Piatek, my paternal grandfather grew up in the small town of Zarowka. His older brother inherited the family farm and there were few job prospects in Galicia. John decided to leave before getting drafted into the Austrian army. His sister was living in Albion and in 1898 he immigrated and joined her. He married Stefania Siebak, the oldest daughter of Adalbert Siebak and Marianna Glowacz. Together they had eleven children, one of which was my father Leo Piatek / Friday. [**Note:** Piatek is the Polish word for the day of the week – Friday]. The Piatek family owned a home at the end of Moore Street in Albion, which was within walking distance to several quarries where John was employed. When my grandfather was furloughed from the quarries at the end of each season, he would go to work at a greenhouse. The facility was operated by the local ketchup factory and provided tomato plants that would get transplanted to the fields in early spring, once the threat of frost had passed. When John was 50 years old, the Orleans County Quarry Company declared bankruptcy (1919) and he went to work full time at the greenhouse and local canning factory.

John Piatek & Stefania Siebak
on Moore St - Albion, NY

My paternal great-grandfather, **Adalbert Siebak** and his second wife Mary Piatek, came from two small villages (Dulcza Wielka and Zarowka) in Galicia, the Austrian controlled portion of southeast Poland. Mary Piatek was also the older sister of my paternal grandfather. Poor economic conditions in their hometowns was the primary reason they decided to immigrate to Albion, NY about 1881. Adalbert found work in the local quarries and at 62 years of age was still listed as a quarryman. At least two of his sons also worked in the quarries.

Adalbert Lukaszyk (aka Lucas) and Mary Szymkowiak were my maternal great-grandparents. They came from Splawie and Krzesinski, two small towns in the Prussian sector on the outskirts of Poznan, Poland. In May 1881 Adalbert immigrated to America, leaving his wife, Mary, who was pregnant with their first child, in Poland. Adalbert initially settled around Detroit Michigan where two of his sisters were living. According to family stories he worked for awhile on the building of a railroad line in Hamtramck. After the birth of their child, Mary immigrated to the US in July of 1882. She and Adalbert settled on Caroline St in the northeast section of Albion, NY. Adalbert, along with his oldest son went to work in the quarries. On the 1915 NY census, Adalbert, who was in his mid 60's was still listed as a quarryman. In the off season he also helped maintain the roads and bridges in Orleans County.

Adalbert Lukaszyk / Lucas
& Mary Szymkowiak
on Caroline St - Albion NY

Three of Mary's siblings, Marcin, Lawrence and Rose and her mother Maria Majewski Szymkowiak also immigrated to Western New York. Like many immigrants of the time, the Lukaszyks were anxious to assimilate into American society. Adalbert simplified his name to George Lucas. It is interesting to note that upon his death his grave was marked by two stones, one with his American name and the second with his original Polish name.

My maternal grandparents, **Anthony Rice** and Rose Lucas/Lukaszyk were both born in the United States shortly after their parents immigrated from Poland. As children they both grew up as neighbors on Caroline Street in the Village of Albion. Anthony Rice worked in the quarries for a number of years. Some time after 1905, around when the Medina Sandstone Company declared bankruptcy, my grandfather decided to change careers and went to work as a sharecropper around Carlton, NY. By 1923, he and my grandmother had saved enough money to purchase a 100 acre dairy farm along the banks of the Erie Canal the end of Orchard Street in the Village of Albion.

Anthony Rice/Reiss & Rose Lucas/Lukaszyk on Orchard St - Albion, NY

As mentioned earlier, only five men of Polish descent lived in Albion, NY at the time of the 1880 US census, and all five were listed as quarrymen. These may be the first five Poles that arrived in Orleans County in the Spring of 1877. [52] They were recruited from another area specifically to work in the quarries. Two of those first Poles were Joseph Daniels (aks Danielewski) and his son Stephan. Baptism records for **Joseph Danielewski** indicate he was born in Wabcz Poland in 1834.[58] Unfortunately, on 29 Aug 1889, the following rather brief obituary appeared in the Medina. NY Tribune newspaper: "Joseph Dan a Polander, and Albion quarryman, was fatally injured on Monday by a premature blast". A more detailed account of the accident also appeared in the Buffalo newspaper.[59] Joseph's son, **Stephan Danielewski** continued to work in the local quarries and went on to marry Rose Szymkowiak, my 2nd great aunt.[60] He also played a prominent role in helping found the Polish Catholic church in Albion.

Stephan Danielewski & Rose Szymkowiak

My material great-grandfather, **Ignatius Reiss / Rice** and my maternal 2nd great-grandfather, **Mathew Kaniecki** first arrived in New York City aboard the SS Vesta on 15 April 1881. Coincidentally, they came from the same small town in Poland as Joseph Danielewski. Most of the residents in Wabcz and the surrounding hamlets were Catholic and attended St. Bartholomew and Ann Church. Joseph Danielewski, his wife Catharina Sosnowska and my ancestors were married and received most of their sacraments in that church. The families likely knew each other before coming to America. I would guess my ancestors first heard about Albion when Joseph Danielewski corresponded with his relatives back in Poland.

Ignacius Reiss/Rice

Around 1885 Ignatius Rice, his family and Mathew Kaniecki traveled back to Poland to visit their relatives. They remained in Poland for about two years and returned to New York aboard the **SS City of Richmond** on 4 April 1887.

[61] On board accompanying my ancestors were about ten families from the Wabcz area that later ended up settling in the Orleans County. Most of the men that immigrated were later listed on the census records as quarrymen. Apparently my

ancestor's favorable reports of the opportunities in Orleans County were enough to convince a number of their friends and neighbors to immigrate to America.

By 1912 the Polish population across Orleans County totaled 1681.[52] In the Village of Albion the Polish community amounted to 175 families and 965 people. They settled mostly in the northeast section of town where they built St. Mary's Assumption Church, a school and an athletic club. The neighborhood was within walking distance to the quarries where many of the men were employed.

An interesting side story: When my Reiss and Kanecki ancestors departed in 1885 for a trip back to their ancestral home, they sailed directly past Bedloe's Island in the middle of NY harbor. When they returned to America in 1887, Lady Liberty was standing tall on the island, welcoming them back home.

Many fine buildings survive as striking reminders of the hard working immigrants from England, Ireland, Germany, Poland and Italy, who took their places in the quarries, and the impact they had on the sandstone industry in Orleans County.

Chapter 5: Sandstone Heritage

There are numerous Medina sandstone structures scattered around the United States and particularly New York State. One of the most impressive examples is in the State Capital building in Albany, NY. The lavish, four story Great Western Staircase was designed by Henry Richardson and built by Isaac Perry. The red sandstone used throughout the project was quarried in Orleans County. In 1897, after laboring for 14 years, craftsmen completed the project at a cost of one million dollars. Today the price tag for the "Million Dollar Staircase" would amount to about $32 million dollars.[62-A&B]

"Million Dollar Staircase"

Having grown up in western New York, I select 19 local examples of Medina sandstone construction. They were chosen in great part because they were familiar fixtures in my life. I used Photoshop software to remove the unsightly power lines from some of the following images.

◊ St Mary's Church ~ Medina, NY

The Irish Catholic Community of Medina, NY began building a church for their congregation in 1898 and it was completed in 1904. St. Mary's Church is a beautiful English Gothic structure constructed with locally quarried brown Medina sandstone. The church was designed by Albert Post, a leading ecclesiastical architect who designed many Catholic churches in Ontario, Canada and the Buffalo area. The primary stone masons for St Mary's were the McDonald Brothers of Lockport, NY and the woodwork was done by the Stokes Brothers of Buffalo. The main building measures 152 ft long and 80 ft wide with a steep gabled, slate roof. Two slate clad steeple towers flank the main entrance. The north steeple is 170 ft and the south steeple measures 100 feet tall. Lavish interior decorations include ornamental molded plaster-work, an intricately carved Carrara marble alter, works by world renowned artists and a historic pipe organ.[63]

◊ St John's Episcopal Church ~ Medina, NY

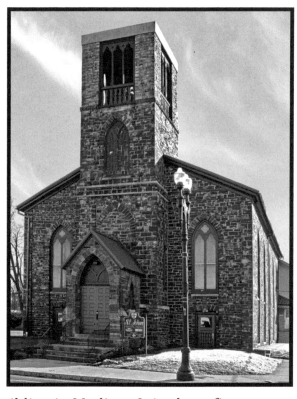

The Episcopal Church of St John is certainly not a towering edifice, but what it lacks in size is more than made up for with its understated charm and historical significance. In 1827 the Mission of St Luke was formed to service the spiritual needs of the Episcopal community of Medina. By 1832 the congregation had grown, changed its name to St John's and made plans to construct a permanent house of worship. John Nixon, one of the parishioners of the parish, was selected as contractor for the project. The church was situated on a slip of land that sits literally in the middle of Church Street. The first services at St Johns were held on Christmas day in 1838, and it is the oldest surviving church building in Medina. It is also a fine example of early red sandstone block construction. The stone for St John's came from near by Oak Orchard Creek. The stone is less well finished than the blocks used in many local churches of later construction.. St John's originally had a spire which was removed after it suffered damage from a cyclone in 1856. A 2100 lb bell still hangs in the stone tower that survived. The interior of the church is simply decorated and well lit by numerous stained glass windows. A fellowship center is located in the basement of the building. Over the years, a small but loyal congregation has worked hard to maintain and restore their cozy place of worship. Given its peculiar location, *Ripley's Believe It Or Not* designated St John's the "Church in the Middle of the Street".[64]

◊ Town Hall ~ Medina, NY

As the Twentieth Century began, Medina, NY had a natural harbor on the Erie Canal and was a burgeoning manufacturing center hosting furniture and clothing plants, foundries and sandstone quarries. The fertile farmland that surrounded Medina also exported large quantities of produce. The town official decided to erect a stately government office building that would reflect the communities bright economic future. In 1908, J.B. McCrady, a noted architect from Buffalo was hired to design the Town Hall. A Richardson Romanesque style structure with rounded arches over the windows using locally quarried red sandstone was erected at the corner of Main and Park Streets in Medina. At some point a sandstone portico was added to the front facade. An extension to the back of the building was also constructed to house the local fire department.[65]

◊ Armory ~ Medina, NY

George Heins was a well known architect responsible for designing numerous churches, municipal buildings and armories. For the Medina Armory he chose a Victorian castle like design complete with octagonal crenellated towers flanking the main entrance and tall narrow windows with iron grates. The imposing 47,928 square foot structure was constructed entirely with locally quarried brown Medina sandstone which complemented the castle motif. Constructed in 1901, the armory was the home to several New York Army National Guard units. Over the years the large interior hall has been used for military drills, athletic events, industrial shows and dress balls. In 1977 the armory was closed and eventually repurposed as the home of the Orleans County YMCA.[66]

◊ Railroad Depot ~ Medina, NY

In 1908 the New York Central Railroad built a new depot on West Main St, to accommodate the brisk passenger traffic on the line. It was a handsome example of a design blending the components of brick and local red Medina sandstone. Prior to 1908, passengers used the adjacent large freight terminal which now houses the Medina Railroad Museum with its extensive collection of local memorabilia and the largest model-train layouts in the country. Fortunately the historic passenger depot was also preserved and has served as the Senior Recreation Center for Western Orleans County for over 50 years. [67-A]

NY Central Depot - Medina NY 1908

My father would have been very familiar with the railroad depot as it was only a few feet from the S A Cook Furniture Factory where he worked for several years as an upholster. My dad and many of the Poles from Albion commuted daily to Medina to manufacture recliner chairs. [67-B]

◊ Pullman Memorial Church ~ Albion, NY

The distinctive church overlooking the Orleans County Courthouse was a generous gift of the industrialist George Pullman in loving memory of his parents who were former residents of Albion. George was the inventor and manufacturer of the Pullman luxury railroad sleeper car. Pullman commissioned Solon Beman, a prolific Chicago architect to design the church. The resulting building has a compact, long, low horizontal profile with attributes of Old English Gothic and Richardson Romanesque architecture. The exterior of the building features reddish brown Medina Sandstone from the local DeGraff & Roberts quarry. The random sized sandstone blocks with a natural rock face were set in an Ashlar pattern with extremely fine mortar joints. Other exterior features include detailed stone carving, a low sloped roof covered with red German fluted tiles, and an octagonal bell tower on the northwest corner of the building. The interior features a Johnson pipe organ, red cedar floors and extensive woodwork of quarter sawed oak. The fifty-six cathedral windows plus another twenty windows circling the central dome were crafted by the Tiffany Studio of New York City. The Pullman Memorial Universalist Church was completed in 1894 and serves as one of the finest examples Medina sandstone construction.[68]

◊ First Presbyterian Church ~ Albion, NY

The First Presbyterian congregation of Albion was established in 1826 and for many years worshiped in the red brick Greek Revival styled building on East State street. When Elizur Hart, a local banker and politician died in 1870, he left a large bequest to the congregation for the construction of a new church. He specifically stipulated in his will that the spire of the new church must be taller than the one on the Baptist church. The well known Rochester, NY architect Andrew Jackson Warner was engaged and produced an English Gothic design for the new building to be situated adjacent to their current house of worship. Locally quarried rustic red Medina sandstone was selected as the primary building material and construction on the church was completed in 1874. A large rose window was positioned above the three entrance doors facing Main Street. Other elements of the new church include pointed arches for doors, interior columns to support the roof of the tall expansive nave, and multiple tall narrow windows to let in lots of light. The single bell tower with steeple and a manse (minister's home) was completed the following year. The steeple rises 175 ft and is still the tallest structure in Orleans County visible for many mile. Located across from each other on the historic Orleans County Courthouse square, the towering First Presbyterian Church and the low compact design of the Pullman Memorial Church are two contrasting examples of architectural artistry and craftsmanship using Medina sandstone.[69]

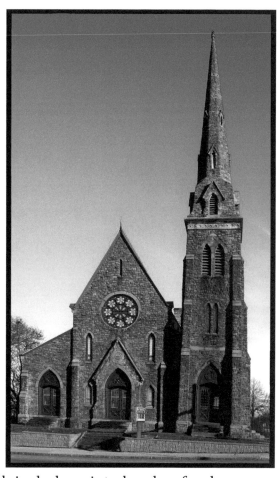

◊ St Joseph's Church ~ Albion, NY

In 1849 the Catholic community of Albion, NY, predominantly Irish immigrants, gathered together to establish a congregation. By 1852 they had constructed a church on North Main St. In 1869 a school, staffed by the Sisters of Mercy was added. In 1896 the congregation had outgrown its facilities and purchased a tract of land at the corner of Park and Main Streets in the Village of Albion.**[70-A&B]** A new St. Joseph's Church was constructed on the site and dedicated in September of 1897, followed by the construction of a new school and lyceum in 1905.**[70-C&D]**

My parents would spend most weekends in Albion visiting relatives. We often attended Sunday services at St Joseph's, which was often referred to as the "Irish Church". I was always impressed by the exterior and interior decor of "St Joe's".

On the outside, St Joseph's Church, like the adjacent lyceum and school were constructed of locally quarried rustic red Medina sandstone. The pointed arches, substantial square bell tower and an interior with a tall ribbed vault supported by a colonnade of marble clad Corinthian style columns are all characteristics of Gothic Revival architecture. The interior walls are clad with a light stone and extensive use of double pane lancet stained glass windows make for a bright cherry interior. The windows were designed by Leo Frohe of Buffalo and depict scenes from the Bible.

By 1919 the church had outgrown its old cemetery located on Brown Road. They needed additional space and the church purchased another plot of land on East Avenue near Mount Albion Cemetery. The land would become known as the "New" St Joseph Cemetery. An attractive little Romanesque style chapel of red Medina sandstone was built by Pasqual DiLaura. It had an octagonal tower, a distinctive red and green terracotta tile roof and stained glass windows designed by the Leo Frohe studio.[70-E] My parents, grandparents and many of my ancestors are buried in this cemetery.

◊ Mount Albion Cemetery ~ Albion, NY

For the first three decades following the settlement of Albion, deceased members of the community were buried in church owned cemeteries or small family plots around the county. In 1843 the village purchased 25 acres of land 1.5 miles east of Albion to be used by the general community as a place to inter deceased loved ones. The tract was a tree covered moraine or mound of sand and gravel deposited by the last receding glacier. Marvin Porter, an engineer who worked on the construction of the Erie Canal, was hired to design the cemetery. He carved out a series of flat terraces for gravesites, added a variety of plantings and laid out a network of roads. Within a year the hilltop was transformed into a park-like setting. Adjacent rolling meadows were added and the cemetery now covers about 100 acres.

In 1875 a chapel was designed and constructed by Charles Diem using locally quarried Medina sandstone. In 1876 a 58 ft high, round sandstone Solders and Sailors Monument commemorating the 453 local men who fell in the defense of the Union during the Civil War was constructed on the highest point of the cemetery. A panoramic view of the countryside can be had from the top of the tower for those hearty souls willing to ascend the 84 steps of the internal spiral staircase. In 1908, as a memorial to local firefighters, a stone spring house and a picturesque lily pond stocked with goldfish were constructed in the northeast corner of the cemetery. Other prominent features of the grounds include an attractive entrance arch and an office building with a maintenance garage across the road from the cemetery, all constructed with locally quarried red Medina sandstone.

Mount Albion was one of the first cemeteries to be listed on the National Register of Historic Places by the United States Department of the Interior. Its 25 miles of winding roads and foot trails, unique landscape and majestic trees serve as one of the first and finest examples of the municipal rural cemeteries in the country. [71-A&B&C]

Many of my relatives were laid to rest at Mount Albion. Today, as in the past it is a quiet serene place for a walk or run. The photo to the left is of my newlywed parents (Leo Friday and Clara Rice), taken atop the Solders and Sailors Memorial Tower, while on an outing in 1936.

◊ St Mary's Church ~ Holley, NY

The Catholic parish of St Mary's in Holley, NY was founded in 1866. Initially the congregation worshiped in a wood framed structure. As the Catholic community in the area grew, a larger facility was needed. A new church was constructed on South Main St of locally quarried red Medina sandstone and dedicated on Christmas Eve 1905. The church has Gothic design characteristics with pointed arches over windows and doors. The main building measures 120 by 34 ft. Two towers, one octagonal and the other a square bell tower are positioned on the right and left front corners of the building. Numerous tall stained glass windows illuminate the interior of the church on a sunny day. St Mary's stands as a proud reminder of the religious fervor, as well as the skill of the hardworking quarrymen of Holley and the surrounding area.[72]

◊ St Rocco's Church ~ Hulberton, NY

The opening of stone quarries about seven miles east of Albion along the canal, attracted immigrants from Italy who ultimately settled around the hamlet of Hulberton, NY. Some of these immigrants were from the small town of Alfedena in the highlands of the Abruzzo region of southern Italy. Many of the immigrants came from a long tradition of working with granite and marble. **"The Road From Alfedena"** a 2006 video by Chris Zinni documented the history and contributions of the Italian-American stonecutters who settled in the Hulberton area.**[73-A,B&C]** In 1906 a number of quarry workers and master stonecutters volunteered their talents and built St Rocco's Catholic Church in Hulberton using local Medina sandstone. The church was renovated in 1976 and the congregation held a festival to celebrate the event. Unfortunately, because of declining enrollment and a shortage of priests, the church was closed and the congregations of St Rocco, St Mark's of Kendall and St Mary's of Holley were merged in 2006. St Rocco's Church and grounds were sold to the Cornerstone Christian Church. However, the St Rocco Italian Festival continues to be held on the banks of the Erie Canal in Hulberton each fall. The popular festival is a celebration of all things Italian, complete with food, music, dancing, boccie tournaments and getting together with family and old friends.

◊ Richmond Memorial Library ~ Batavia, NY

The Richmond Memorial Library is located on part of the old Richmond Estate on Ross Street in Batavia, NY. It was commissioned in 1889 by local philanthropist Mary E. Richmond, wife of Dean Richmond a railroad magnet and first vice president of the New York Central Railroad. The library was designed by Rochester architect James Culter in the Richardsonian Romanesque style. The original building is a one-and-a-half-story T-shaped structure constructed with gray Medina sandstone in a random Ashlar pattern, trimmed with red Albion sandstone. The building has a steeply pitched gabled roof. The main entrance is off-center with ornate carvings around the door. On the right corner of the building is an octagonal tower with a conical shaped copper roof. The library was renovated and doubled in size in 1970.**[74A&B]**

The Richmond was my library while I was growing up. Before the days of the internet, I spent many hours there researching school term papers and reports.

◊ First Baptist Church ~ Batavia, NY

The First Baptist congregation in Batavia was founded in 1834. Their current church is located on 306 East Main St, directly across the street from St Joseph's, the church I attended. Pierce and Dockstader of Elmira were the architects and John Shaefer was the contractor for the project. The church was built and dedicated in 1891. Much like the Richmond Library, this church exhibits many of the classic design characteristics of Richardson Romanesque architecture. The solid masonry walls utilize natural rock faced gray Medina sandstone accented with red Medina sandstone. The large circular stain glass window facing Main St is surrounded by a wide rounded arch. The main roof and both entrances feature front facing gabled roof lines. The design also features a large square bell tower and a prominent tall circular tower that serves as a chimney.[75]

◊ St Mary's Church ~ Batavia, NY

I grew up in Batavia, NY in Genesee County, just south of Orleans County. During the early history of Genesee County the religious needs of the Catholic community were served by missionaries and St Joseph Church in Batavia. In 1904, as the community grew, Sacred Heart of Jesus Parish was established to meet the needs of Polish immigrants and St. Mary's parish was also formed to meet the needs of Catholics living in the western part of the city. In 1908, St. Anthony of Padua Parish was added to address the needs of Italian immigrants in the area. In 1906 St. Mary's Parish began construction of a new church for their congregation. They engaged John Copeland from Buffalo to design a Gothic style dwelling which included a crenellated bell tower, pointed arched windows and small side projections resembling buttresses. The church was built on Ellicott St by a local contractor John Pickert using reddish brown, natural rock faced Medina sandstone blocks. The stained glass windows were provided by Leo P. Frohe, from Buffalo and illustrate scenes from the lives of Jesus and his mother Mary.[76A&B]

◊ St. Bernard Seminary ~ Rochester, NY

Saint Bernard's is a historic former Roman Catholic seminary complex located on Lake Avenue in Rochester, NY. It was situated on a 20 acre plot overlooking the Genesee River. Bernard J. McQuaid, the first bishop of the Diocese of Rochester was the driving force behind the establishment of the seminary. The campus is composed of a group of 4 interconnected buildings designed by noted Rochester architects Warner & Brockett and built between 1891 and 1908. The four buildings are the Center or Main Building (1891–1893), the Chapel (1891–1893), the South Building or Philosophy Hall (1900), and the North Building or Theology Hall (1908). All four buildings share a Victorian Gothic style of architecture with stone walls and brick trim. The buildings make extensive use of a distinctive variety of Medina sandstone quarried at nearby Hanford's Landing in the Genesee gorge adjacent to Kodak Industrial Park in Rochester, NY. This sandstone is red with large, very distinct white inclusions.

The seminary graduated over 2700 priests, but because of declining numbers of seminarians, it moved its campus in 1981. The buildings were owned for a time by Eastman Kodak Co. They were sold to a private developer and converted into a senior citizen residential complex.[77]

◊ Holy Sepulchre Cemetery ~ Rochester, NY

In the early part of the 19th century there were five Catholic churches in the City of Rochester and each had a small cemetery. By the time Bernard McQuaid was appointed bishop of the Catholic Diocese of Rochester, some of the small cemeteries were running out of room. The bishop decided to build one large central burial ground that would be easier to manage and offer facilities that would meet the needs of the Catholic Community for years to come. He selected a large tract of land just north of the City of Rochester on the banks of the Genesee River bisected by what is now Lake Avenue. The vision Bishop McQuaid had was to develop a peaceful, garden like cemetery. In 1871 he enlisted F. R. Elliott of Cleveland, Ohio to design the first sections of what came to be called Holy Sepulchre Cemetery. Noted local horticulturalists and nurserymen such as Patrick Barry and George Ellwanger selected and supplied many of the plantings. A garden expert from Belgium, Pierre Meisch, was hired to construct a road system and to act as the first superintendent of the facility. Gatehouses resembling a castle, (see photo below) were constructed of red Medina sandstone, and located on both sides of the eastern entrance of the cemetery. They served as the initial offices for the facility. A 1.4 mile long, 4 foot high wall of red Medina sandstone was also built on both sides of Lake Ave to define the boundaries of the property.

Proceeding through the eastern entrance, just past the gate houses, is the magnificent centerpiece of the cemetery. The All Souls Chapel and bell tower was constructed over a period of 10 years and were finally completed in 1886. The local architect, Andrew Jackson designed a beautiful Early English Gothic committal chapel with red Medina sandstone, a steeply pitched slate roof and two rose windows facing the front gate. The interior of the

chapel was lavishly finished with stained glass windows and there is also a below ground vault. A stunning 110 ft tall square bell tower, also of red Medina sandstone, is adjacent to the chapel and was topped with two round extensions. The soothing sounds of tower bell and carillon can be heard throughout the cemetery.

In 1919 Alling DeForest, a local landscape architect was hired to develop a vacant section of the cemetery along the Genesee River gorge with rolling hills and a pond. He divided the single pond into three interconnected bodies of water. Three sandstone bridges where constructed and fountains were added to the ponds creating a peaceful, serene landscape known as Trinity Lake.

Holy Sepulchre now covers about 322 acres and is the largest, most active of Rochester cemeteries. The grounds are impeccably maintained and many additions and modifications have been made over the years. The cemetery and the adjacent grounds of St Bernard's Seminary are situated just north of the Kodak Research Laboratory where I worked for many years. They were often locations for a relaxing lunch hour walk.[78]

◊ Charlotte Lighthouse ~ Rochester, NY

The Charlotte–Genesee Lighthouse was built in 1822 on a bluff 85 ft above Lake Ontario at the mouth of the Genesee River. It stands adjacent to the Port of Rochester and Ontario Beach Park. The 40 foot tall octagonal tower was constructed of rough cut Medina sandstone and later painted white for better visibility. The tower is 24 feet around at the base and 11 ft around at the top with a copper roof. 10 whale oil lamps originally served to illuminate the lighthouse which were replaced in 1858 with a Fresnel lens. The brick and sandstone lighthouse keeper's dwelling was built adjacent to the old tower in 1863. The lighthouse was decommissioned in1881 and a signal light was placed at the end of the harbor pier. The lighthouse keeper and later the commander of the local Coast Guard station occupied the keeper's home. In 1994 the lighthouse and keeper's dwelling were deeded to the City of Rochester and today serve as a historical museum and offers regular tours. [79]

◊ Hamlin Beach, NY State Park

Hamlin Beach State Park is located along the shore of Lake Ontario at the very northwest corner of Monroe County adjacent to Orleans County. It has been a popular outdoor recreational destination for many years with a beautiful one mile long sandy beach, large campground, picnic areas and ten miles of hiking trails. The tract of land was acquired by Monroe County in 1929 and has an interesting history. From 1935 to 1941 the park was home to a Civilian Conservation Camp where the men lived on site and worked to develop the park. Medina sandstone was widely used to construct numerous shelters, walls restrooms, walkways breakwalls, bridges and roads. From 1944 to 1946 the park also served as a German prisoner of war camp. Today the park covers 1287 acres and attracts about 300,000 visitors each year.**[80-A]**

Bathhouse at Hamlin Beach State Park

Construction of the 35 mile long Lake Ontario State Parkway began in the late 1940's. The highway follows the shoreline of the lake and serves to connect the city of Rochester and rural towns with the parks and recreational facilities along the shores of Lake Ontario. Local Medina sandstone was used to construct a number of bridges and overpasses all along the parkway. **[80-B]**

Lake Ontario Parkway - E. Manitou Rd. Overpass

◊ Dziadziu's (Grandpa) Milkhouse ~ Albion, NY

Tony Rice - my grandfather

Not all Medina sandstone buildings were designed by a famous architect or decorated with fine art work. There are many simple, utilitarian structures scattered around Orleans County constructed with locally quarried sandstone. The humble structure pictured above is located on my grandparent's dairy farm on Orchard St in the Village of Albion. Every morning and again in early evening my grandfather would milk his small herd of cows by hand and pour the liquid into cans. He would store the cans in his milkhouse in a large refrigeration chest filled with water until a truck would come to collect them. That milkhouse was always a cool place to visit on a hot summer day and it stands as a reminder of how blessed I was that my grandparents were such an important part of my life.

The story of the sandstone quarries of Orleans County and the waves of immigrants that toiled in them are an important chapter in the history of Western New York. The nineteen buildings presented here are only a small fraction of the many examples of structures that stand as visible reminders of the hard work, talent, courage and endurance of the immigrants that left their homelands and came to America with their dreams for a better life. They brought with them a strong work ethic, new customs, new foods and new ideas that made this country a richer more vibrant place. We inherit an America built on a foundation of their efforts.

~ References ~

References (Chapter 1);

[1-A] Physical Geology (EEN1110 – Tulane University) by S. Nelson 2018 https://www.tulane.edu/~sanelson/eens1110/sedrx.htm
[1-B] Sedimentary Rock – Wikipedia
[https://en.wikipedia.org/wiki/Sedimentary_rock]
[1-C] 1-B Aggregate Geology and Classification (E117 Concrete and Technology & Code 4)

[2-A] Porosity Prediction in Quartzose Sandstones as a Function of Time, Temperature, Depth, Stylolite Frequency, and Hydrocarbon Saturation - AAPG Bulletin by P. Bjørkum, etal. -1998.
[2-B] Predicting Porosity through Simulating Sandstone Compaction and Quartz Cementation - Geology; AAPG Bulletin by R. Lander & O. Walderhaug (1999)

[3-A] Global geologic maps are tectonic speedometers—Rates of rock cycling from area-age frequencies by B. H. Wilkinson, etal. GSA Bulletin (2009) 121 (5-6): 760–779.

[3-B] Petrogenesis of Metamorphic Rocks by K. Bucher, R Grapes (2011), p24

[4] Composition of Earth's Continental Crust as Inferred From the Compositions of Impact Melt Thaw Sheets- David A. Kring, Lunar and Planetary Science XXVIII (1997)

[5-A] Ordovician and Silurian sea–water chemistry, sea level, and climate-by Munnecke etal -2010

[5-B] Geologic Structure and Occurrence of Gas in Parts of Southwestern, NY- US Dept Interior Bulletin 899b, Part 2-G. Richardson (1941), p77

[6] Rainbow of Rocks by M. Chen and W. Parrey 2002]

[7-A] Landmarks of Orleans County, NY by I.Signor (1894) p8-14

[7-B] Medina Sandstone Quarry Prospectus by Church and Ryan 12 Dec 1899 - Lee-Whedon Memorial Library, Medina, NY

[7-C] New York State Museum V. 2. No.10, Building Stone in New York, - J. Smock (1890), p220,

[8]   Plate Tectonics, Encyclopedia of Geology, by R.C. Searle, p340 (2005)

[9-A] A Walk to the Appalachians by J. Renten (2014)

[9-B] A Geological Study of Pennsylvania 4th ed.by J. Burnes (2002)

[9-C] Geology of New York - 2nd Edition, Y. Isachsen (2000)

[9-D] Birth of the Mountains by Sandra Clark (2001) & etal , 2000

[9-E] Map at top of p10 derived from the Great Appalachian Valley Map (Greatvalley-map.jpg) from Wikimeda Common (public domaine)

[9-F] Diagram at bottom of p10: derived from two sources – US Geological Survey Ground Water Atles [HA-730-L 1997 Fig. 9] and [H-730-M Fig 1]

[10-A] Description of Niagara Quadrant by E. Kindle (1912)

[10-B] Geology of New York; Part IV, Survey of the fourth geological district by J. Hall (1843), p34-54

[11-A] Map drawn from – Geologic Map of New York – 1970 (, NYS Museum and Science services – Series 15)

[11-B] Revised Stratigraphy and Correlations of the Niagaran Provincial Series (Medina, Clinton, and Lockport Groups) in the Type Area of Western New York By Carlton E (1995)

[11-C] Landmarks of Orleans County, NY by I.Signor (1894) p152-161

[11-D] Our Stone Quarries - Medina Tribune – 19 Oct 1892

[11-E] Journal Register – Medina, NY: Only memories remain of Sandstone Industries by J. Hudnut 3 Feb 1955

[12-A] Stratigaphy of the Genesee River Gorge at Rochester, NY SGA by T. Grasso (1973)

[12-B] Rochester History V54 No. 4 (1992) Geology and Industrial History of the Rochester Gorge by T. Grasso

[13-A] Elements of Geology ~ by J.LaConte (1904) p235

[13-B] Methods of Quarrying and Dressing, Smithsonian Institute (1886), p 310

[14-A] Geologic Structure and Occurrence of Gas in Parts of Southwestern New York - US Dept Interior Bulletin 899b by G. Richardson (1941)

[14-B] In Search of a Silurian Total Petroleum System in the Appalachian Basin of New York, Ohio, Pennsylvania, and West Virginia – by R. Ryder, etal. (2014)

[14-C] The Devonian Marcellus Shale and Millboro Shale - by D Soedar (2014)

Chapter 2 References:

[15-A] www.Ancestry.com > Search Public Trees >John Ryan (Birth: 13 Feb 1801) > John Ryan Family Tree

[15-B] www.Ancestry.com > Search Public Trees > Leo Piatek Friday (Birth: 11 Apr 1909) > Friday Family Tree

[15-C] Orleans County Genealogical Records – www.orleans.nygenweb.net

[16] Landmarks of Orleans County, NY by I.Signor (1894) p326

[17] John Ryan-Obituary, Rochester Democrat & Chronicle (21 Dec 1896)

[18] Bulletin of, NY State Museum – V 2, No 10 - Building Stone of, NY by J Smock p261-265

[19-A] Bird Coler Obituary - Daily News- New York- 3 June 1941

[19-B] Stone – An Illustrated Magazine V24 (1902): p113 & 261

[19-C] Medina Quarry Co - Buffalo Review - 19 Mar 1902

[19-D] Titles Pass in Formation of Trust -Buffalo Evening News – 22 Mar1902

[19-E] Cole Managed Medina Quarry Co Deficit $111,000 - - 20Sept1906

[19-F] Medina-Orleans Quarry History -Brooklyn Daily Eagle: 19Sept1906, and Part of Reorganization Farce - 21 Sept 1906

[19-G] Lease and Sale of Medina Property Denounced as Looting and Fraud -Brooklyn Daily News-22Sept1906

[19-H] Wall Street Men Gain Quarries -Brooklyn Daily Eagle -23Sept1906

[19-I] Decision Involves $1M void mortgage -Buffalo Evening News-29Sept1909

[19-J] Coler Accused of Paving Graft - Chicago Tribune 1906- 29Aug1906
[19-K] Coler Co Were Shareholders in Orleans County Quarry Co-Brooklyn Daily Eagle- 4Sept1906
[19-L] Wall Street Men Gain Quarries -Brooklyn Daily -23Sept1906
[19-M] Coler v Eagle Lible Suit Summation 1 - Brooklyn Daily Eagle - 2Mar1909
[19-N] Coler v Eagle Lible Suit Summation 2 - Brooklyn Daily Eagle - 3Mar1909
[19-O] Coler v Eagle Lible Case - Coler Lost - Brooklyn Daily Eagle - 14Jan1910
[19-P] Medina's Tainted trail Makes Buffalo Scandal -Brooklyn Dail Eagle -12Oct1906
[20-A] Image of quarry ad on p 20: A Souvenir Book of the Village of Albion - Orleans County, NY, 1905 , p 90
[20-B] Quote: A Souvenir Book of the Village of Albion - Orleans County, NY, 1905, p 152
[20-C] Image of Canal Toweline on p 30 courtesy of Orleans County Historical Society
[20-D] Canals For A Nation: The Canal Era in the United States, 1790-1860 by R. E. SHAW (1990)
[20-E] Images of steam boat on canal on p 31 courtesy: Steve Hicks - .albionalumni.org/chevrons/alb/cas2.html Albion Canal, P2, image 10
[20-F] Image of Medina Train Depot -on p 31 courtesy;: Steve Hicks - www.albionalumni.org/chevrons/alb/cas2.html Medina, P3, # 17
[20-G] Map on p32 courtesy; *"1826-1926 One Hundred years of Railroading"*: New York Central Railroad (Publicity Department pamphlet) by C. F. Carter also   New York Central Railroad 1831-1915 by A. H. Smith (1916)
[20-H] Photo of Hulberton Sandstone on Canal on p33 courtesy; Town of Holley and Murray Historian (Marsha DePhilips)
[20-I] Major Canals Built in the 19th Century, American Northeast
The Geography of Transport Systems, 5th Ed,Jean-Paul Rodrigue (2020), - also see [https://transportgeography.org/?page_id=1128]
[20-J] Erie Canal National Heritage Corridor website [www.eriecanalway.org/]

[21-A] Many Italians In Stone Quarries in Albion: Democrat & Chronicle – Rochester, NY 2 May 1905
[21-B] Sandstone Industry Dwindles - Rochester Democrat & Chronicle by W. Monacelli - 13 Jul 1958

[21-C] Sandstone Quarries had a Golden Age: The Daily News On Line – by Matt Ballard (24 Sept 2018)

[22] War Losses -USA- International Encyclopedia of the First World War V1 by CR Byerly -2014

[23A] Global Mortality of the 1918-1920 "Spanish" Influenza Pandemic by N. Johnson & J Mueller - Bulletin of the History of Medicine Vol. 76, No. 1 (2002), pp. 105-115

[23-B] The Influenza of 1918 by M. Humphreys, Evolution, Medicine, and Public Health [2018] pp. 219–229

[23-C] Americas Wars - Dept of Veterans Affairs and Dept of Defense

[24-A] Moody's Manual of Railroads and Corporation Securities, Vol. 2, (1920) p 1286 (foreclosure sale)

[24-B] History of Medina Sandstone History 1894-1964 by Pasquele DiLaura – Hoag Library - Albion, NY (archive collection).

[24-C] Mineral Resources of US 1922 by GF Loughlin

[25-A] Model T Ford: The Car That Changed the World Hardcover – 1994 by Bruce W. McCalley

[25-B] Federal Aid Road Act of 1916: Building The Foundation – by R. Weingroff, Public Roads Magazine - Vol. 60 No. 1, 1996

[25-C] Bitulithic Hot Mix Patent # US727505 – F. Warren

[26] Sears, Roeback & Co - Concrete Machinery Catalog (1925) p 12,16&17

[27] New Product - Medina Red sandstone - Stone 2004

[28] Mining and Quarry Industry of New York State Report of Operations and Production (1914) by D.H. Newland

[29] Municipal Engineering - Vol 25 (1903) p319 - Medina Sandstone Pavements in Brooklyn, NY

[30-A] Municipal Journal & Public Works - Vol 14 (1912) p 194 >197 - Medina Block Payment by E.Fisher (50 years experience)

[30-B] Reclaimed stone photo on p25 courtesy: W. Farmer Jr. - Catenary Construction Co – Rochester, NY, (www.catenaryconstruction.com)

[31-A] Story of the New York State Canals by R. Finch 1925

[31-B] History of the Barge Canal of, NY State by N Whitford (1921)

[32] Erie Canal Fact Narrative (provided by Camillius Erie Canal Museum)

[33] Marco Paul on the Erie Canal by Jacob Abbot (1852) (fictional story but historically accurate depiction of life on the Canal)

[34] Landmarks of Orleans County, NY by I.Signor (1894) p67

[35] Water-ways from the Oceans to the lakes by t. Clarke -Scribners Magazine, Vol. XIX, no. 15, 1896, p. 103-113

[36-A] Supreme Court: Frank Calonna vs State of New York (1907), claims #17,17, 17337
[36-B] Quarryman Case Upheld - Democrat & Chronicle - Rochester, NY - 9 May 1930
[36-C] Quarry Owners in Orleans Demand Redress - Buffalo Currier - 6 Apr 1919
[36-D] Albion - M Ryan Sues for Damages - Buffalo Times, 26 May 1920

[37] Orleans County Quarry Co vs State of, NY - 3May1916

Chapter 3 References:

[38-A] History of Medina Sandstone History 1894-1964 by Pasquele DiLaura – Hoag Library - Albion, NY (archive collection – Delia Robinson).
[38-B] Oral History Project – Orleans County Historical Society – Marcus Phillips Interview (1978) – Hoag Library - Albion, NY (archive collection).
[38-C] Sandstone Quarrying in the US (Dept of Interior - Bulletin 124-17) by O. Bowles (1917)
[38-D] Elements of Geology: A Text-book for Colleges and for the General Reader By Joseph LeConte(1905)
[38E] Handbook of Rock Excavation Methods & Cost - H P Gillette (1916)

[39-A] BUILDING STONES OF ONTARIO –Part 1- Report 14 -D. Hweitt (1964)
[39-B] Sandstone Quarries of Apostle Island by K. Eckert - 1985

[40] Thesis on the Blasting and Quarrying of Stone - John Burgoyne 1895 [https://books.google.com/books/about/A_Treatise_on_the_Blasting_and_Quarrying.html?id=dKdJAAAAIAAJ]   (Google e-books)

[41-A] Image of double jackers on p36: Early Stone Cutters in Western Missouri, Poplar Heights Farms, Hand Drilling Stone (2005) (public domain)
[41-B] Methods of Quarrying and Dressing - The Collection of Building and Ornamental Stones in the US National Museum (1886) by G Merrill, p 313

[41-C] Plug and Feather photo on p 37 courtesy Peggy B. Perazzo, - Stone Quarries and Beyond Continues www.quarriesandbeyond.org (Memorial and Cemetery Review 1923)

[41-D] The Chronicle V59, #2, June 2006, p36: Tools and Machinery of the Granite Industry, Part II by Paul Wood

[42-A] The Chronicle V59, #3, Sept 2006, p87: Tools and Machinery of the Granite Industry, Part II by Paul Wood [see Figure 22: Trimming a block]

[42-B] photo of quarry workers on p38 courtesy of *OrleansHub.com* (Immigrants were critical in growth of local sandstone industry - M Ballard, -22 Sept 2018 also courtesy Medina Sandstone Society (Dave Miller)

[42-C] Guy Derrick photo on p39 courtesy: Kansas Geological Survey, Bulletin 142, pt. 2, 1960, Kansas Building Stone by H Risser (public domain)

[43-A] Gang saw photos on p40; Manufacturer and Builder Journal V16 1884 p277-278

[43-B] Gang saw photos on p40; The Georgia Marble Co.*An Illustrated Magazine*, Vol. XLVI, No. 3, March, 1925,

[43-C] The Chronicle V59, #4, Dec 2006, p133: Tools and Machinery of the Granite Industry, Part III by Paul Wood [Fig 7&8 gang saw]

[43-D] Kansas Geological Survey, Bulletin 142, pt. 2, 1960 , Kansas Building Stone by H Risser

[43-E] Scientific American V66 #6 p89 - Steam Stone Works – 6Feb1892

[44-A] Hand tool image on p41 courtesy: Brenner & Lay - Tools (catalog - Chicago, Ill) ca 1900 courtesy Peggy B. Perazzo, Stone Quarries and Beyond Continues   [www.https://quarriesandbeyond.org/index.html]]

[44-B] Stone Cutters Tool Box, STONE V38, 1917, p24

[45-A], NYS Museum 56th Annual Report - 1902 - Economic Geology of Western, NY by I Bishop r45

[45-B] Souvenir Book of the Village of Albion-Orleans County, NY, 1905, p90

[46-A] Photo of Medina Culvert on p43 courtesy: Orleans County Historical Society also see   Steam Donkey and Drill - Manufacturer and Builder Journal V17 1885 p107-8

[46-B] Photo of guy derricks on p43 courtesy: Elements of Geology: A Textbook for Colleges and for the General Reader By Joseph LeConte(1905),-fig. 197, p 236

[46-C] 2 Steam Tractor photos on p 44 courtesy: Image courtesy of Orleans County Historical Society

[46-D] Stone crusher photo on p 44 and two quarry photos on p 50 courtesy: A Souvenir Book of the Village of Albion - Orleans County, NY, 1905,
[46-E] Blake Crusher images on p 45 & 46 courtesy of 911 Metallurgy Corp. [www.911metallurgist.com/blog/blake-jaw-crusher]
[46-F] Image of Trommel Screen on p 46 courtesy Wikimedia Commons Ⓒ
[46-G] Photo of crusher crew on p 47 courtesy: Hoag Library – Albion, NY
NOTE: I found a second vary revealing image **(copyrited)** of a similar crusher operation. Rock Crusher, Webster Springs, W. Va. https://wvhistoryonview.org/catalog/015419

[47-A], NYS Museum 56th Annual Report - 1902 - Economic Geology of Western, NY by I Bishop r45
[47-B] Souvenir Book of the Village of Albion-Orleans County, NY, 1905, p90
[47-C] Drawing of channeler on p45 courtesy: Sullivan Machine Co – 95 year commemorative publication courtesy Peggy B. Perazzo, Stone Quarries and Beyond Continues [https://quarriesandbeyond.org/index.html]
[47-D] Scientific American, Vol. XXVIII, No. 13, 1873 - Improved Stone Quarrying Machine
[47-E] Channeling Pressure Stone-mine & Quarry (1912) Sullivan Steam Drill

[48-A] Drawing of steam drill on p47 courtesy: Methods of Quarrying and Dressing - The Collection of Building and Ornamental Stones in the US National Museum (1886) by G Merrill, p 320
[48-B] The Chronicle V59, #2, June 2006, p44: Tools and Machinery of the Granite Industry, by Paul Wood
[48-C] Early Stone Cutters in Western Missouri, Poplar Heights Farms, Hand Drilling Stone > Tools > Hand Drilling Stone
https://www.stone.poplarheightsfarm.org/machine_drilling.HTM
[48-D] Quarrying with Hammer Drills – Mine and Quarry Vol. VI, No.1, Aut 1911, pp. 537-540
[48-E] Steam Drill Applications: Manufacturer & Builders Journal V7 1885 p59-61

[49-A] Photo of Albion Quarry with Channeler on p 46 courtesy; A Souvenir Book of the Village of Albion - Orleans County, NY, 1905, p 90
[49-B] Photo of Steam shovel at top of p49 courtesy :A Souvenir Book of the Village of Albion - Orleans County, NY, 1905, p 89
[49-C] Photo of Steam shovel at bottom of p49 courtesy; Steve Hicks - www.albionalumni.org/chevrons/alb/cas2.html Albion Canal, p2, image 12
[49-D] Two Goodrich Quarry photos on p 50 courtesy: A Souvenir Book of the Village of Albion - Orleans County, NY, 1905, p 88 & p 92

[49-E] Photo of Horan Quarry on p 51: Bulletin of Geology Society of America -V16 -1905 by H Fairchild , p51

[49-F] Photo of McCormik Quarry on bottom of p51: courtesy of Orleans County Historical Society

[49-G] Photo of Cor nell Quarry at top of p52: Bulletin of Geology Society of America -V16 -1905 by H Fairchild , p51

[49-H] Photo of Hulberton quarry on bottom of p52 courtesy of Town of Holley and Murray Historian (Marsha DePhilips)

[49-I] Photo of quarry workers by pond on p 53 courtesy: Tom Rivers - www.OrleansHub.com/vintage, Hulberton Quarry Worksers 12 Apr 2013

[49-J] Photo of Farrand Quarry on top of p 54: courtesy of Orleans County Historical Society also Medina Sandstone Society (Dave Miller)

[49-K] Journel Register – Medina, NY: Only memories remain of Sandstone Industries by J. Hudnut 3 Feb 1955

[49-L] Photo of Crusher Plant on bottom of p 54 courtesy of Town of Holley and Murray Historian (Marsha DePhilips)

[49-M] OBrine Quarry photo p 55 courtesy www.OrleansHub.com/vintage, Quarry workers Role with Hamlin Beach Parkway by Tom Rivers 10 Oct 2014 also see Journal Register – Medina, 5 Feb 1985

[49-N] Photo of OBrine Quarry top of p 56 courtesy of Town of Holley and Murray Historian (Marsha DePhilips)

[49-O] Four photo of Vincent quarry on p 57-59: courtesy of Town of Holley and Murray Historian (Marsha DePhilips)

Chapter 4 References:

[50-A] Gabriola's millstone quarry - *SHALE* No.19, Nov 2008 by J. Gehlbach
[50-B] Monthly labor review - U.S. Depart. of Labor, v.49 1939 Jul-Dec p60
[50-C] United States. Bureau of Labor Statistics. "Aug1960," *Employment and Earnings* (August 1960) p29

[51-A], NYS Bureau of Inspections Annual Report -60B -1906
[51-B] STEREOSCOPIC X-RAY EXAMINATION OF SANDSTONE QUARRY WORKERS by D Kindel and E Hayhurst (1926)

[52] Immigrants in Industries, - Part 24, p533-538 (The Immigration Commission) 1912

[53] 1880 US Census (Orleans County-Albion, NY), [orleans.nygenweb.net/]

[54-A] Irish and German Immigrants of the Nineteenth Century: Hardships, Improvements, and Success by A. Tagore (2014)
[54-B] Irish Immigration to America, 1630 to 1921 by Dr. C. Shannon
[54-C] GERMAN AMERICANS IMMIGRATION by D. Cunz (1952)

[55-A] THE CAUSES OF POLISH IMMIGRATION TO THE UNITED STATES by Sister Lucille-Polish American Studies, Vol. 8, No. 3/4 (1951), p. 85-91
[55-B] Library of Congress "Nation of Polonia" (www.loc.gov)

[56] History of Italian Immigration by A. Molner

[57-A] Albion Seven Sisters - Rochester Democrat & Chronicle -12Aug1928
[57-B] Images of America: Albion -A. Townsend (2005) ISBN 0-7385-3929-5, p56.

[58] Wabcz (Chełmno / Culm), Bydgoszcz, Poland - Church records; 1760-1890 [14 Mar 1843] (LDS microfilm Roll# 162328 p 100 & 285)

[59] Accident to a Quarryman – Buffalo Weekly Express 28 Aug 1889.

[60] Albion, NY St Joseph Church records, 1856-1916 (24 Sept 1884) (LDS microfilm roll # 1378511)

[61] SS City of Richmond [*1887; Arrival: New York, New York, USA*; Microfilm Serial: *M237, 1820-1897*; Line: *38*; List Number: *349*]

Chapter 5 References:

[62-A] Foundation Stone of Western, NY by D.Nichols - Conservationist, NYSDEC (Jan 1990) p32-37
[62-B] Photo on p72 courtesy: Anaconda Standard – Montana (30 Mar 1911)

[63] St Mary's Roman Catholic Church – Median, NY by C. Bush (2010)

[64] We feel it is a duty to Medina to keep it up' By Tom Rivers, Editor Posted 7 July 2014 at 12:00 am [www.orleanscountyhub.com

[65 Medina's Historic City Hall: An Enduring Piece of WNY's Architectural Heritage - Chris Busch Nov17, 2012: http://orleanscounty.wgrz.com/news/70959

[66] "Medina, NY Armory" ~ Wikipedia also , NYS Division of Military & Naval Affairs: Military History http://dmna.ny.gov/historic/armories/Medina.html

[67-A] Image of Medina depot on p77; courtesy Steve Hicks - www.albionalumni.org/chevrons/alb/cas2.html Medina, P4, image 22

[67-B] Cook's Automatic Chair (3 Apr 2007) [https://www.kovels.com/collectors-questions/cooks-automatic-chair.html] also US Patent #667162 by G A Bowen (29 Jan 1901)

[68] Pullman Memorial Church -, NYTimes 1895

[69] Landmark Presbyterian Church as Testament to 19th Century Prosperity – Orleans.com – Mathew Ballard – 20 Aug 2016

[70-A] Souvenir Book of the Village of Albion-Orleans County, NY, 1905 p46
[70-B] Setting the Record Straight_ William Staffords Spiteful Sale to St Joseph's bt Mathew Ballard (2016)
[70-C] Democrat & Chronicle – Rochester, NY [29 June & 31 Dec 1896] and Catholic Union & Times -Buffalo, NY 23 Sept 1897
[70-D] The Catholic Church in the United States of America: Undertaken to Celebrate Golden Jublie of Pope Pius X – Vol III (1914) p484
[70-E] A Cemetery's Journey–The Daily News by Mathew Ballard (27 Feb 1916)

[71-A] A Souvenir Book of the Village of Albion - Orleans County, NY, 1905 p119 to 121
[71-B] Signor, Isaac S. (1894). *Landmarks of Orleans County, New York*. Syracuse, NY: D. Mason & Company. pp. 272–273.
[71-C] Orleans County built Soldiers & Sailors Monument to honor local Civil War Sacrifice by Mathew Ballard 25 May 2018), OrleansHub.com (Tom Rivers)

[72] The Catholic Church in the United States of America: Undertaken to Celebrate The Golden Jublie of Pope Pius X – Vol III (1914) p503-4.

[73-A] Italian-American Traditions in Western New York by T. M. Reynolds (1999), http://memory.loc.gov/diglib/legacies/loc.afc.afc-legacies.200003410/
[73-B] **"The Road From Alfedena"** documentary video (2006)
by Chris Zinni PhD, Dept Anthropology, Brockport SUNY
[73-C] Immigrant Tales by   - Batavia Daily News, 22 Jul 2006]

[74-A] Richmond Library Dedication – Batavia Daily News 13 Mar 1889
[74-B] www.batavialibrary.org/about > History of the Library
[75] Batavia is home to 2 grand churches made of Medina sandstone by Tom Ricers (28 Oct 2016),   www.OrleansHub.com
[76-A] St Mary's Roman Catholic Church :
www.co.genesee.ny.us/departments/history/batavia_walking_tour.php

[76-B] The Catholic Church in the United States of America: Undertaken to Celebrate The Golden Jublie of Pope Pius X – Vol III (1914) p487

[77] Whatever Happened To St Bernard's Seminary by Alan Morrell Democrat & Chronicle – Rochester, NY (20 Sept 2014)

[78] HOLY SEPULCHRE CEMETERY by E Vogt – Rochester History - V67, 2005

[79] Charlotte-Genesee Lighthouse:
[https://www.lighthousefriends.com/light.asp?ID=304] and [www.seathelights.com/ny/charlotte.html]

[80-A] Hamlin Beach State Park History (Wikipedia):
https://en.wikipedia.org/wiki/Hamlin_Beach_State_Park
[80-B] Lake Ontario State Parkway (Wikipedia):
https://en.wikipedia.org/wiki/Lake_Ontario_State_Parkway

◊ Hard and soft cover versions of this book can be purchased directly from my website: **www.jimfriday.com** >> **go to Gallary 7** or **e-mail** me directly at photos.jimfriday@gmail.com

◊ This book is published by **IngramSpark®** and available through retail vendors;   ISBN: 9781087942827 (hard cover)
ISBN: 9781087942520 (soft cover)
Library of Congress Control Number: 2021901672

CPSIA information can be obtained
at www.ICGtesting.com
Printed in the USA
BVHW022045140221
600124BV00004B/33